IMAGES
of Modern America

Lost Steel Plants of the Monongahela River Valley

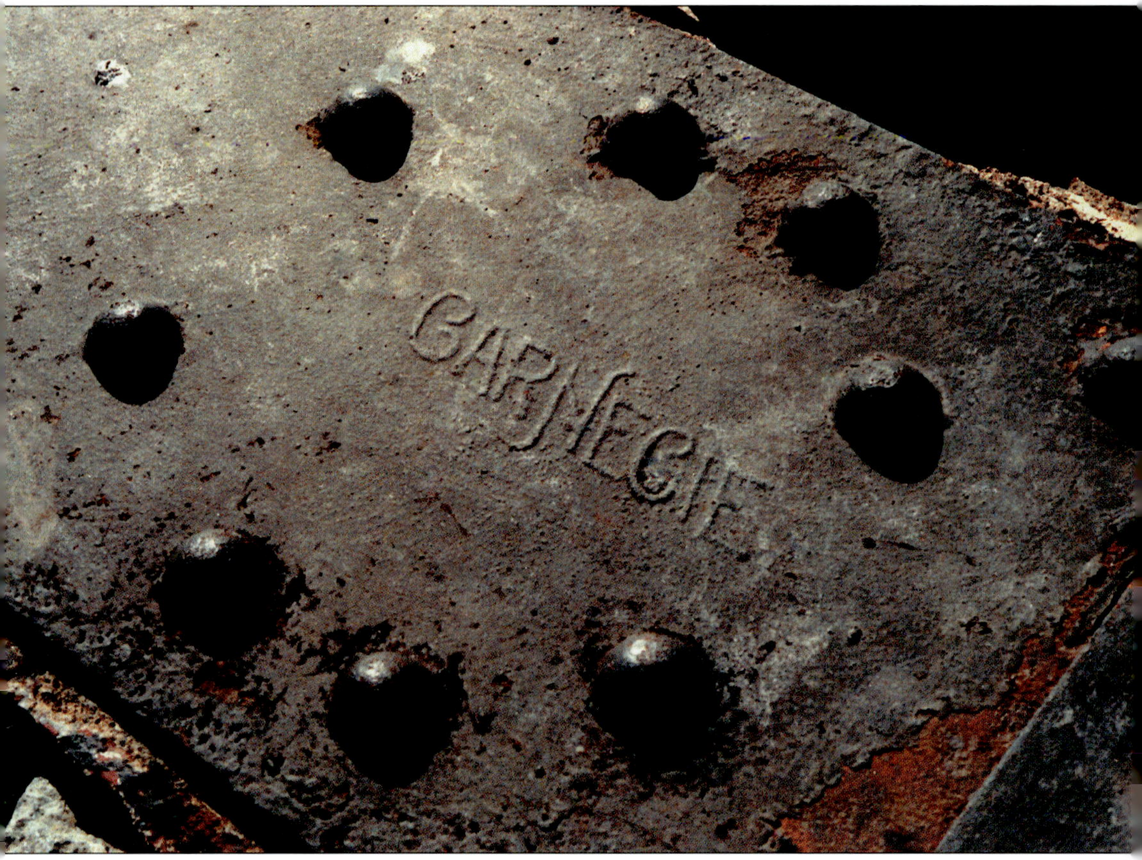

Walking along the old Pennsylvania Railroad main line tracks just east of Pittsburgh, one will come upon this Carnegie steel mark on a train overpass, built in 1914. (Author's collection.)

On the Front Cover: Clockwise from top left:
Edgar Thomson at night (Author's collection; see page 14), Open-Hearth Building No. 5 at Homestead (Author's collection; see page 24), abandoned train engine at Homestead (Author's collection; see page 31), Carrie Furnace, as seen from downriver (Author's collection; see page 44), Carrie Furnace's hot metal bridge (Author's collection; see page 86)

On the Back Cover: From left to right:
Homestead (Author's collection; see page 26), Homestead's forge shop (Author's collection; see page 35), Homestead workers' locker (Author's collection; see page 33)

IMAGES
of Modern America

Lost Steel Plants of the Monongahela River Valley

Robert S. Dorsett

ARCADIA
PUBLISHING

ISBN 978-1-4671-3466-8

Published by Arcadia Publishing
Charleston, South Carolina

Printed in the United States of America

Library of Congress Control Number: 2015941080

For all general information, please contact Arcadia Publishing:
Telephone 843-853-2070
Fax 843-853-0044
E-mail sales@arcadiapublishing.com
For customer service and orders:
Toll-Free 1-888-313-2665

Visit us on the Internet at www.arcadiapublishing.com

To my parents, Robert and Mary Dorsett; to my friend Jason Majewski for his encouragement; and to iron and steel workers everywhere

Contents

Acknowledgments

When I started taking photographs of steel plants in 1984, I was using a 35-millimeter single-lens reflex camera with a 28-millimeter, a 50-millimeter, and a 135-millimeter lens. I used Kodak Gold color film, 200 Advanced Photo System (APS), with 24 exposures per roll.

These old photographs have a "look" that is impossible to duplicate with a digital camera. Most of these images are from high-resolution scans of my original color prints. In 2009, I bought a digital single-lens reflex (SLR) camera. Some of the Carrie Furnace photographs are from digital images.

I am thankful that I had the opportunity and the inspiration to make a photographic record of what was being lost to the changing times. I was too late to take photographs of the steel plants in operation, but I did document some of the demolition. These sites are now mixed use technology parks and shopping malls. The steel industry could rightly be said to have built and then nearly destroyed the region economically and environmentally.

Only photographs and memories remain to show our children what was once there. Some young people have expressed to me their interest in these old photographs of what was lost. They want to know what it looked like, what was there before the shopping malls. They wish they could have seen what the steel plants looked like before their lives began. Their interest has been key to my making this book. I hope that you find my photographs of interest and can recognize the beauty of the mills and blast furnaces that I have tried to capture for posterity.

All photographs appear courtesy of the author.

INTRODUCTION

The Monongahela River's name is derived from the Lenape Indian word Menaonkihela, which means "Where banks cave and erode." The Shawnee, Seneca, and Iroquois first inhabited this area of Southwestern Pennsylvania. The first European, a French fur trader by the name of Robert de la Salle, came up the Ohio River in 1669.

The British and French vied for control of the region. In 1755, the British sent troops led by General Braddock and George Washington to attack Fort Duquesne. They were defeated by the French and Indians at Braddock's field, now the site of the Edgar Thomson steel plant. On November 27, 1758, the British, under General Forbes, took control of the abandoned Fort Duquesne (named for the French governor of New France) at the confluence of the Monongahela, Allegheny, and Ohio Rivers and renamed it Pittsburgh after William Pitt, the British secretary of state.

By 1768, descendants of William Penn took control of the Pittsburgh area from the Native American Six Nations through the Treaty of Fort Stanwix.

In July 1797, George Washington put down an insurgency against the whiskey tax with 13,000 state militiamen at Braddock's field.

In the 1790s, Chestnut Ridge in Fayette County had iron plantations, which smelted iron for the early settlers. The area was also blessed with large reserves of coal, which had been used and exported since the beginning of the 19th century.

Pittsburgh in the 1840s was already known as the "Iron City." Allegheny County was the biggest iron-producing region in the country. Rolling mills would roll the crude iron into bars, which were then taken and hammered into finished products like nails and horseshoes used by settlers going West.

At this time, Allegheny County was also the foremost producer of bituminous coal in the United States and an important manufacturer of steam engines used on the rivers for transportation into the interior of the country. By the 1860s, half of the boats on the Mississippi River system were made in Western Pennsylvania.

Sheffield Steel from England was widely used in America as the preferred metal. When the Civil War broke out, English steel became hard to get; thereafter, Pittsburgh steel became the standard. As steel supplanted iron as the primary metal, the region became internationally recognized as the heart of the steel industry in the United States.

By the late 1860s, using the Bessemer converter and open-hearth furnace to make steel, the Carnegie Steel Company was the largest manufacturer of pig iron, steel rails, and coke in the world.

Railroads became an important method of transportation used by the steel industry. Narrow-gauge rail lines were used inside, and broad-gauge railroads were used between the plants, both to bring in raw materials and to transport finished products to a growing nation. The Union Railroad was created by Carnegie in 1876 to handle the various switching tasks within each of eight major steel plants, including the Edgar Thomson, Duquesne, and Homestead plants. The

Port Perry Bridge over the Monongahela River opened to traffic on June 30, 1898, and the Carrie Furnace Bridge opened to hot metal traffic on December 31, 1900. Jones and Laughlin's hot metal bridge also opened in 1900.

The history of steel making in Western Pennsylvania is a story that touches so many lives it is hard to overestimate its impact. Workers from all over Europe, including English, Scottish, Slovaks, Romanians, Hungarians, Poles, and Serbians, came to Pittsburgh. They came seeking good jobs working in the steel plants along the Monongahela River.

They made the steel that built the Panama Canal, the Empire State Building, and the Gateway Arch in St. Louis; the armor plate steel that built the US and Russian navies; and the steel for the tanks and bombs used in two world wars. The fire and smoke from the industry lit up the night skies so brightly that airplane pilots flying from New York to Chicago used them to navigate by.

Labor at the US Steel Homestead Works was represented by the Amalgamated Association of Iron and Steel Workers (AA), formed in 1876. They negotiated wage scales, helped regulate working hours, and improved working conditions.

Management at Carnegie Steel wanted to break the union. They wanted more profits and lower wages for fewer workers. This all led to the "battle of Homestead" on July 6, 1892.

During the lockout, headquarters for the AA were in the Columbia Hotel at 623 East Eighth Avenue and Heisel Street. From the roof, people at AA could look out into the Homestead plant, and newspaper reporters used the telegraph at the hotel to send news of the lockout to the rest of the country. Today, this building holds the offices of the Rivers of Steel Heritage Corp.

H.C. Frick, chairman of the Carnegie Steel Company, hired Pinkerton detectives to oust the workers from the US Steel Homestead Works. When the detectives landed in two barges at the pump house, the tow boat withdrew, and they were left stranded. Shooting began, and there were casualties on both sides. The Pinkerton men surrendered at 5:00 p.m. The state sent in 8,000 militiamen to end the strike.

As the economy went, so did the steel industry. In 1911, Pittsburgh was producing half of the nation's steel. The Great Depression hit the region hard, but preparation for World War II saw the building of new blast furnaces. US Steel reached its highest employment in 1943 with 340,000 workers. In 1953, production peaked at more than 35 million tons per annum. With the dumping of cheap foreign steel in the 1970s due to price fixing from state-owned steel producers and the comparatively higher cost of labor in the United States, the industries' aging steel plants along the Monongahela River began their decline.

Following the 1981–1982 recession, the plants laid off 153,000 workers. By 1990, seven of the nine major steel plants were closed. Edgar Thomson is one of the few plants still in service. Current operations include two blast furnaces (Nos. 1 and 3) and a continuous caster installed in 1992 to convert steel directly into slabs. In 1995, the Edgar Thomson plant was designated a Historic Landmark by American Society for Metals (ASM) International. In 2005, the plant produced 2.8 million tons of steel, 28 percent of US Steel's domestic production. Today, US Steel is the eighth-largest producer of steel in the world.

One

Edgar Thomson Plant
Carnegie's First

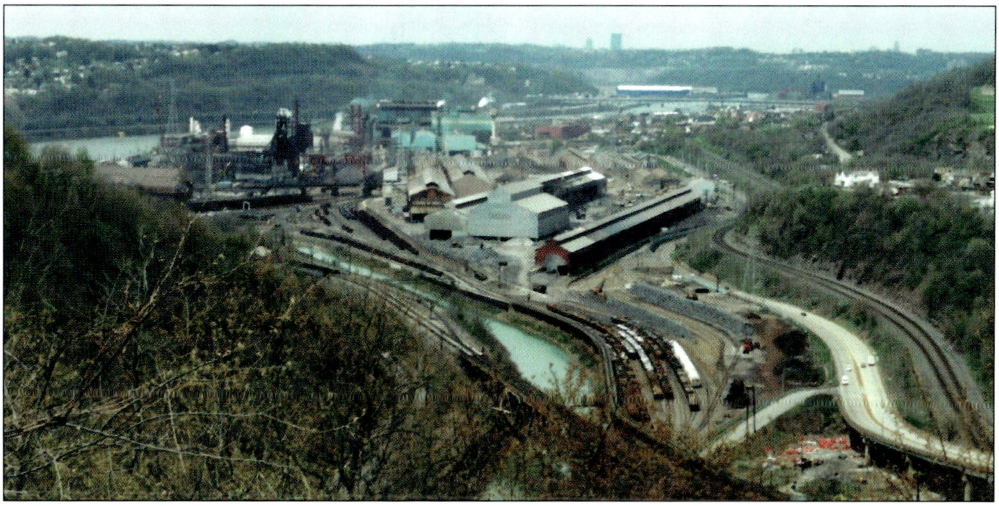

This photograph shows the Edgar Thomson plant, located where George Washington fought against the French and Indians while serving under General Braddock in 1755. The Baltimore & Ohio (B&O) and the Pennsylvania Railroads both have main line tracks passing through here. Edgar Thomson made steel rails for both railroads. Westinghouse Air Break Company is located up the Turtle Creek valley to the right. Note the US Steel Building on the horizon in downtown Pittsburgh.

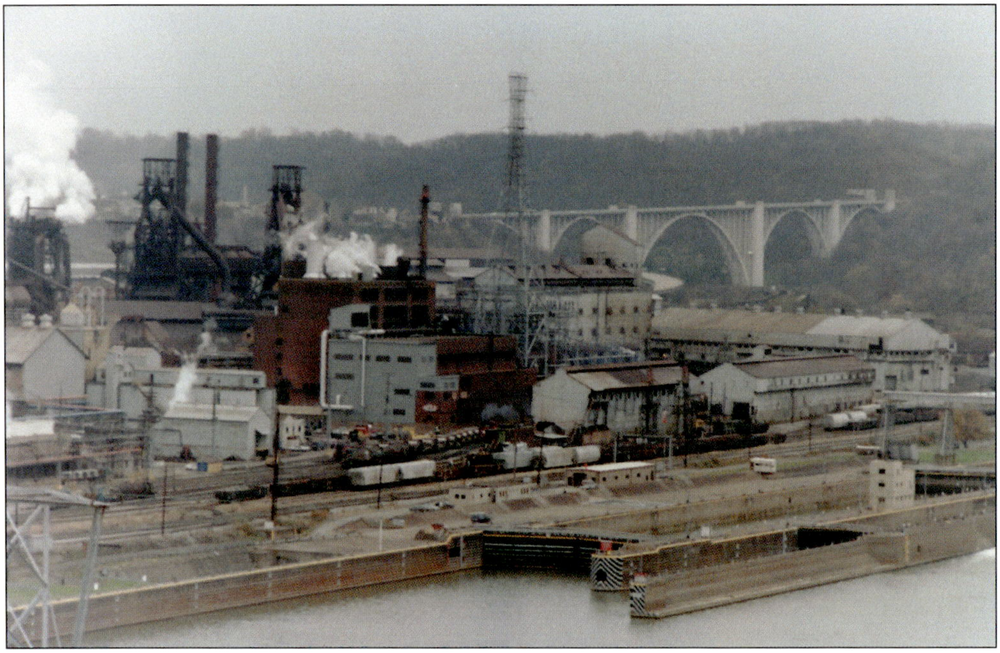

This photograph shows, from left to right, Blast Furnaces Nos. 2 and 1 as well as other parts of the Edgar Thomson plant, which opened on August 22, 1875, as viewed from across the river. Lock No. 2 on the Monongahela River is visible in the foreground, and George Westinghouse Bridge, built in 1932, which carries US Route 30 over Turtle Creek, is pictured in the distance.

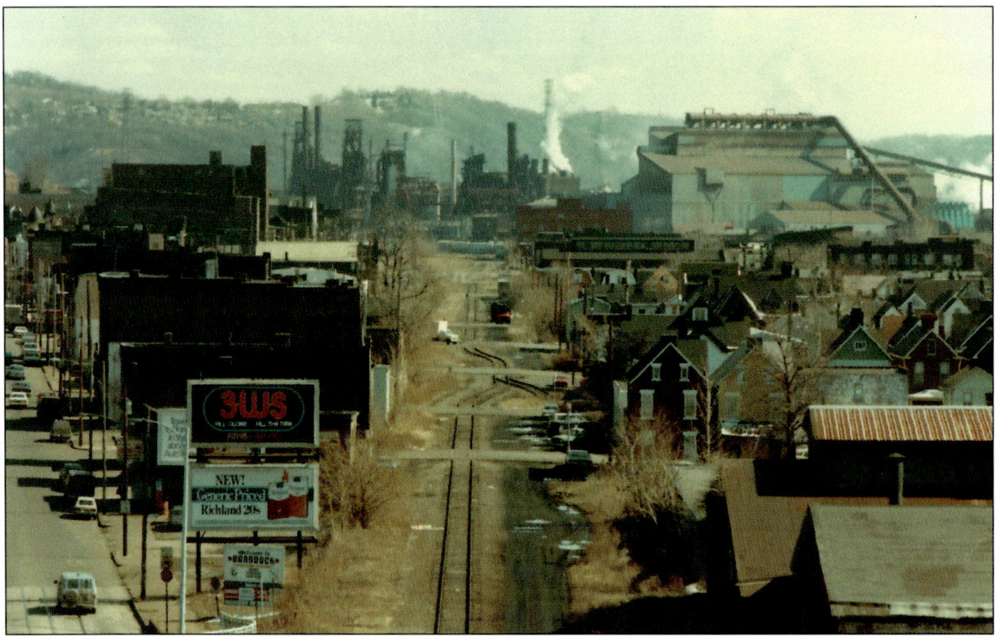

This picture, taken from the George Rankin Bridge, is of the Edgar Thomson plant and the once-thriving town of Braddock, Pennsylvania—one of the towns that was decimated by the collapse of the steel industry. In 1950, the population was 16,488. By 2010, it had declined to 2,159. However, it continues to evolve, with some industry and hopes for the future.

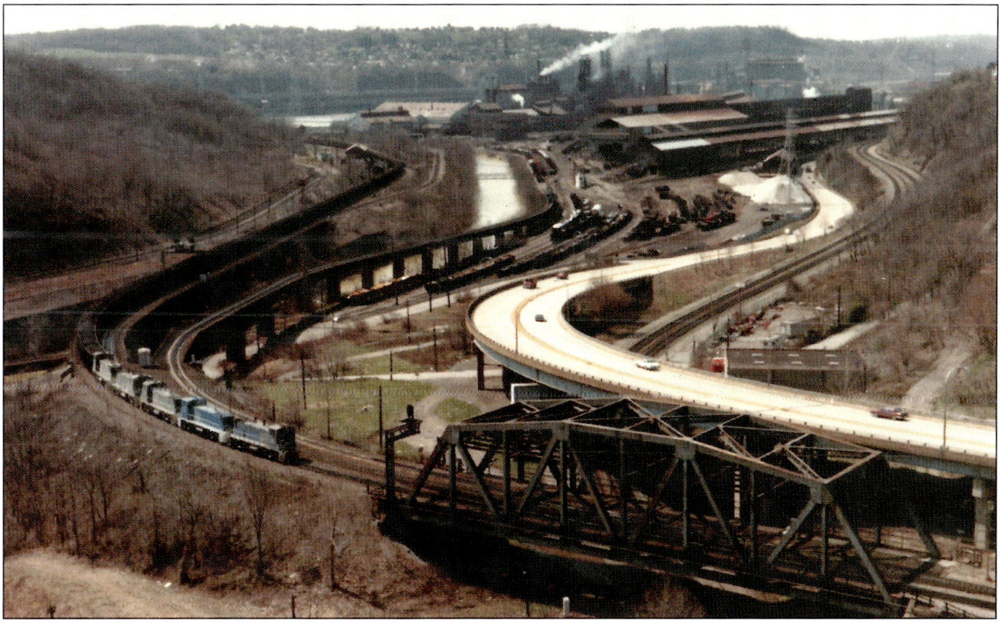

In this photograph is a Union Railroad train taking its load of coal past the Edgar Thomson plant, heading east up the Turtle Creek valley to the Bessemer and Lake Erie yards. From there, the coal is taken up to Lake Erie and loaded onto ships bound for Canada. The train tracks that snake along the base of the hill to the right were once the Pennsylvania Railroads main line tracks.

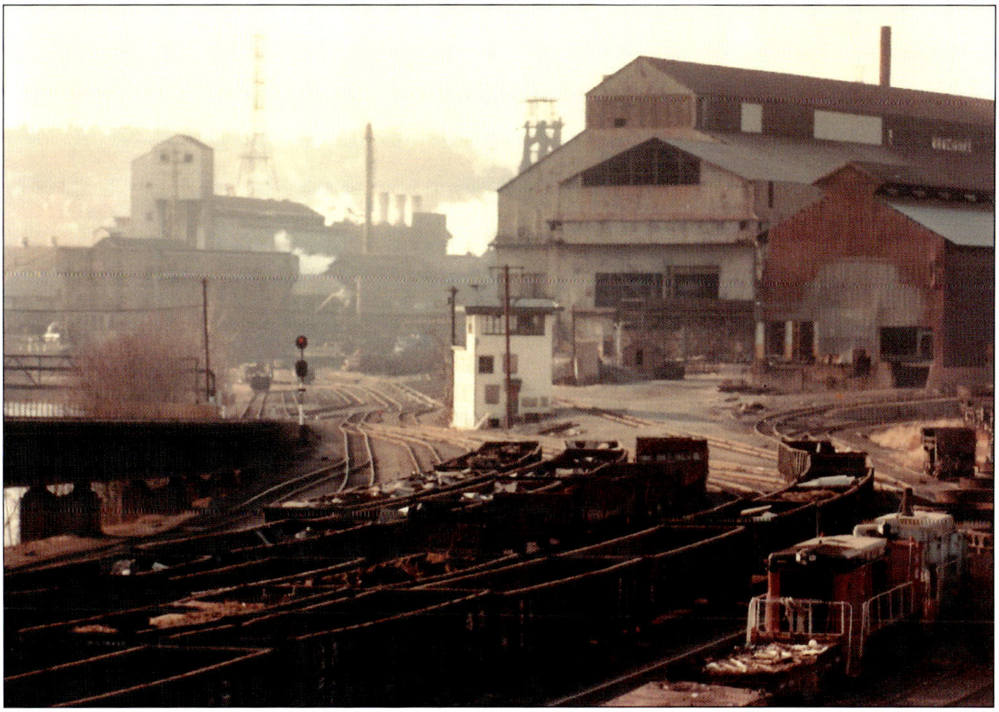

Here is a closer view of the train yard on the east side of the Edgar Thomson plant. The train ramp on the left, going down into the mill, is called the low-grade line. The white building, center frame, was the valley yard office. It was removed in the mid-1990s.

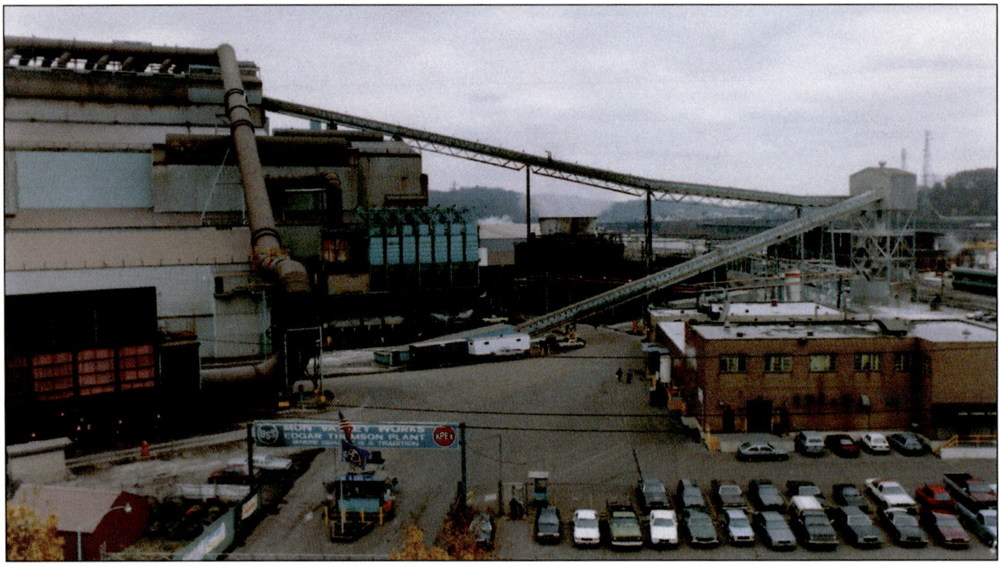

Workers are seen entering the plant at Gate No. 4, along the western side of the plant. Taken from the roof of the Talbot Towers Public Housing Building, now removed and replaced with a gas storage tank farm, this eastward-facing photograph shows the big, blue Basic Oxygen Process (BOP) building (left) with dark-blue gas cleaners on the side. A conveyer belt and Administration Building are on the right.

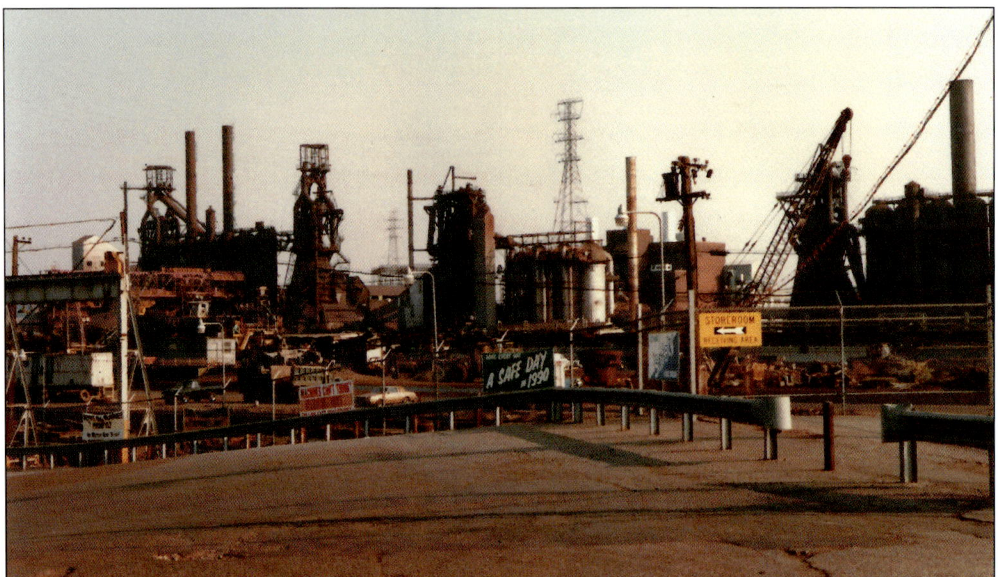

Pictured in 1990 at Gate No. 2 are four blast furnaces. They are, from left to right, Blast Furnaces Nos. 1 and 2, No. 3 by itself, and No. 4 on the right. Nos. 1 and 2, "high-tops," were built by the Defense Plant Corp. around 1940. Blast Furnace No. 2 was removed to make space for the slag pit system used by the newly overhauled and renovated Blast Furnace No. 3.

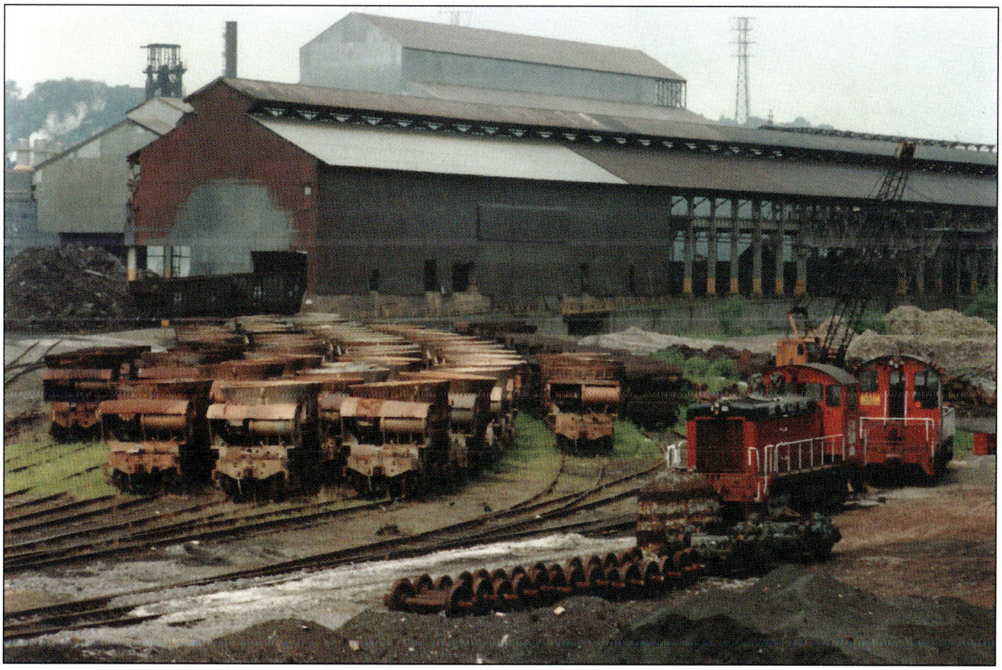

Looking into the Edgar Thomson plant, one sees ladle cars lined up in the eastern train yard. On the right, two remotely controlled switcher engines wait to move them to be filled with molten iron and then move them into the ladle metallurgy facility. At the ladle metallurgy facility, the molten iron is refined into steel, moved to the dual strand continuous caster, and then transported to the Irvin works to be rolled into products.

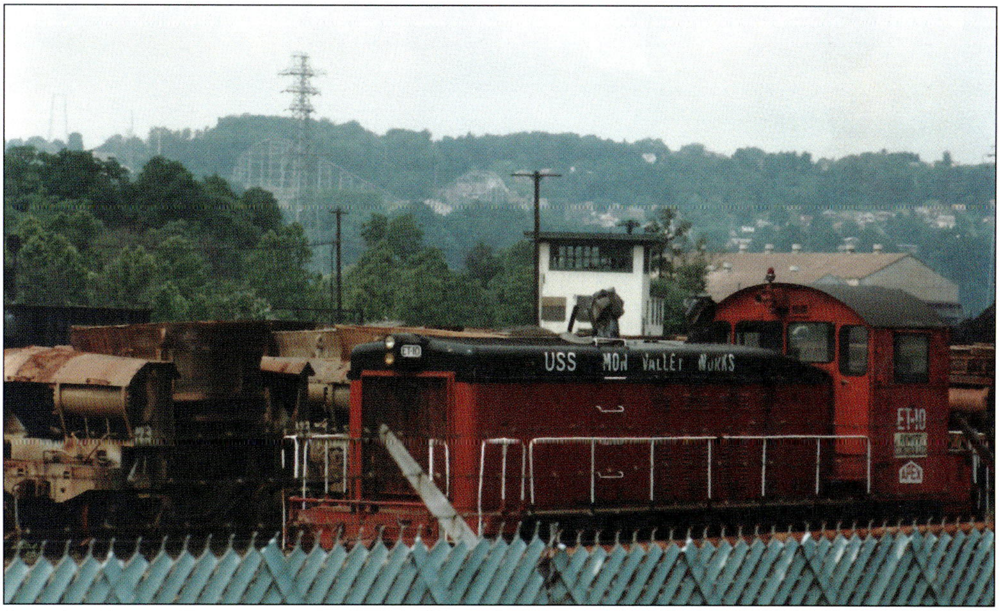

Switcher Engine ET-10 is at US Steel's Edgar Thomson plant; the engine is remotely controlled. In the background, a roller coaster at Kennywood amusement park is seen across the river. People working at the plant could hear the excited yells of the people on the rides, and people at the park could see the fire and smoke of the mills, each wondering about the other.

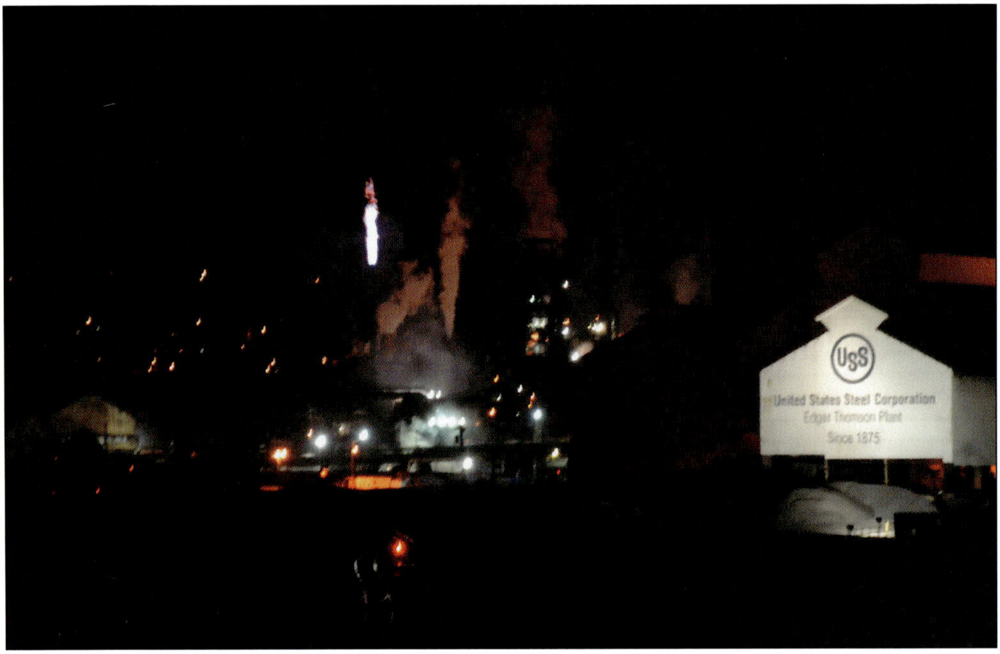

A nighttime summer view of the Edgar Thomson plant, with its blast furnace working all through the night, shows its majesty, mystery, beauty, and power. The flare tower is shooting bright-blue flame high up into the night sky. Smoke and lights and sound and smell, the plant is alive with the power that makes steel.

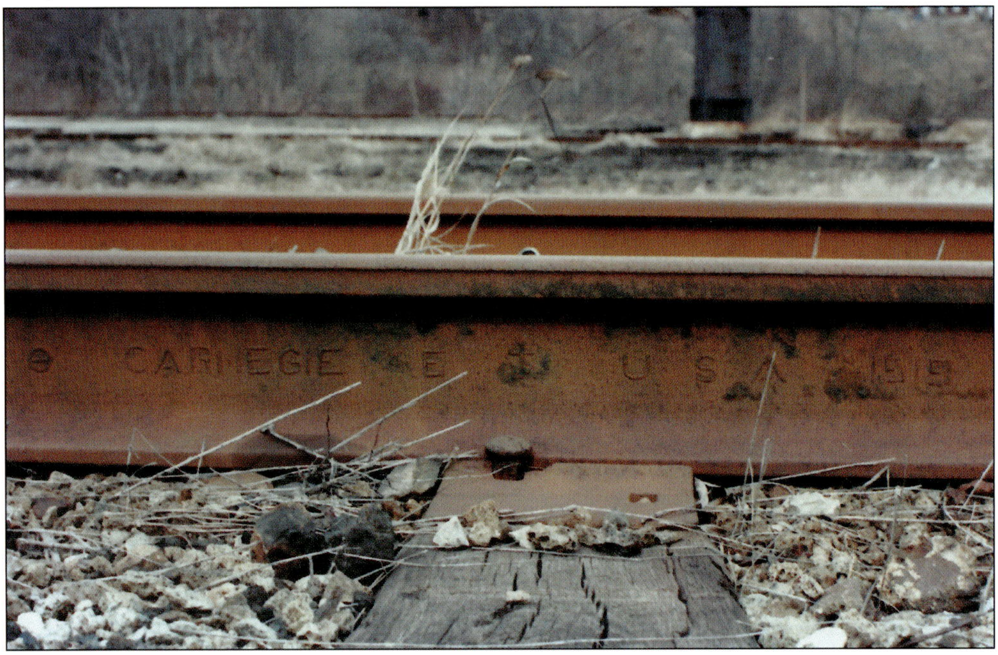

A train rail was found at the old Pennsylvania Railroads Pitcairn Yards, named for Robert Pitcairn, superintendent of the Pittsburgh Division of the Pennsylvania Railroad. "Carnegie ET USA 1919" is written on its side. Today, the facility is the Norfolk Southern Pitcairn Intermodal Terminal. This rail has seen a lot of traffic in 70 years.

Two

Duquesne Plant
One-Time Rival

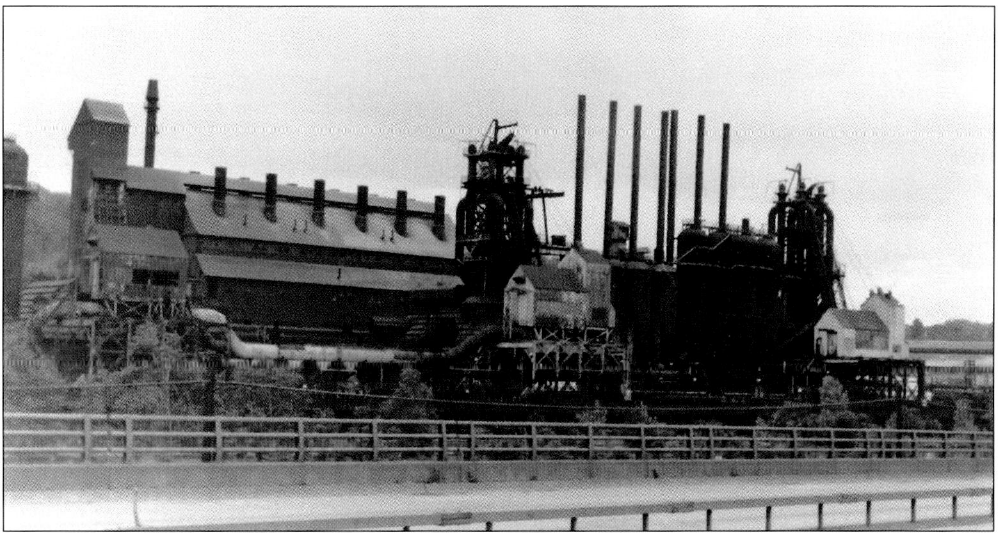

In this northwestern view, State Route 837 passes before, from left to right, the US Steel Duquesne plant's Blast Furnaces Nos. 3 and 4 in Duquesne, Pennsylvania. The red-roofed buildings in front of the furnaces are the hoist houses, which used Otis elevator motors to pull cables that lifted skip cars full of materials, like coke, limestone, and iron ore, to the top of the furnaces.

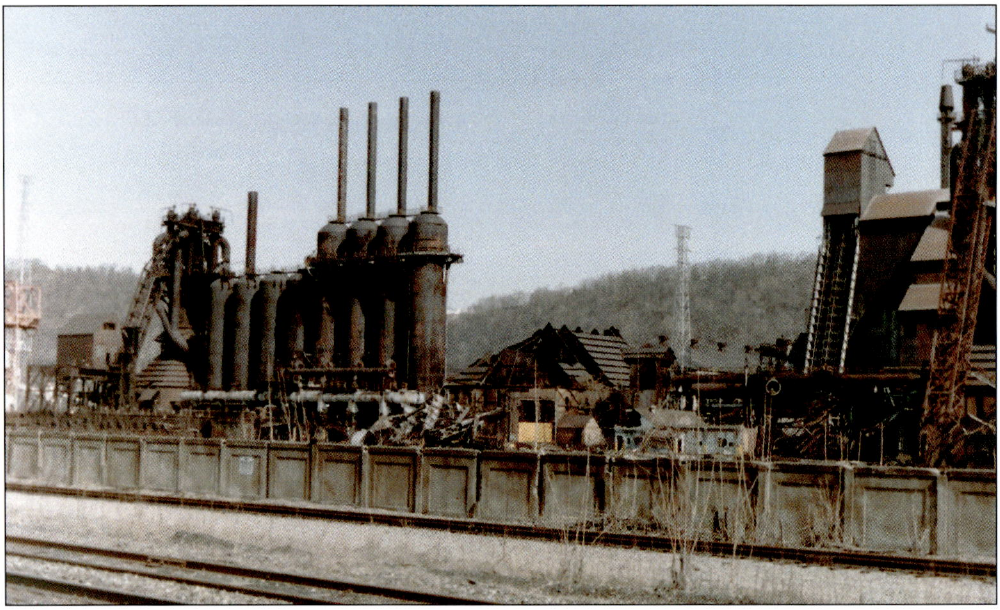

This northeastern view of the Duquesne plant shows Blast Furnace No. 1 on the left, built in 1902. Already demolished to the right is Blast Furnace No. 2, built in 1903. Next on the right is Blast Furnace No. 3. Each furnace had four Cowper hot-blast stoves and draft stacks. The Carnegie Steel Company bought the Duquesne plant in 1890.

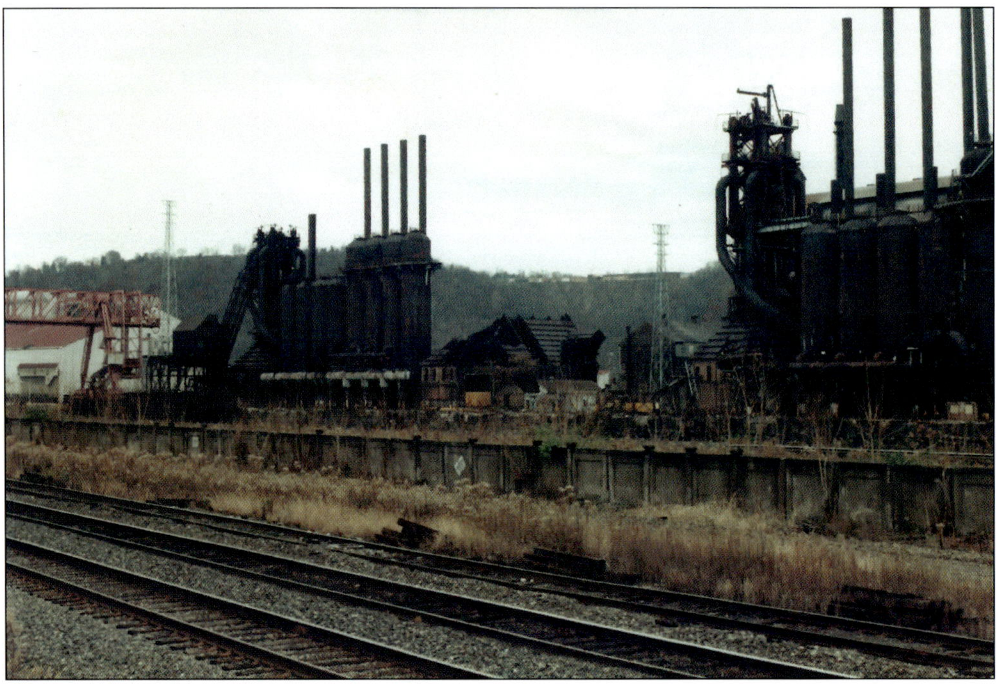

From left to right are Duquesne plant's Blast Furnaces Nos. 1 and 3. These blast furnaces each have four Cowper hot-blast stoves, each stove with its own draft stack. The railroad tracks going past the furnaces were operated by Conrail until 1999.

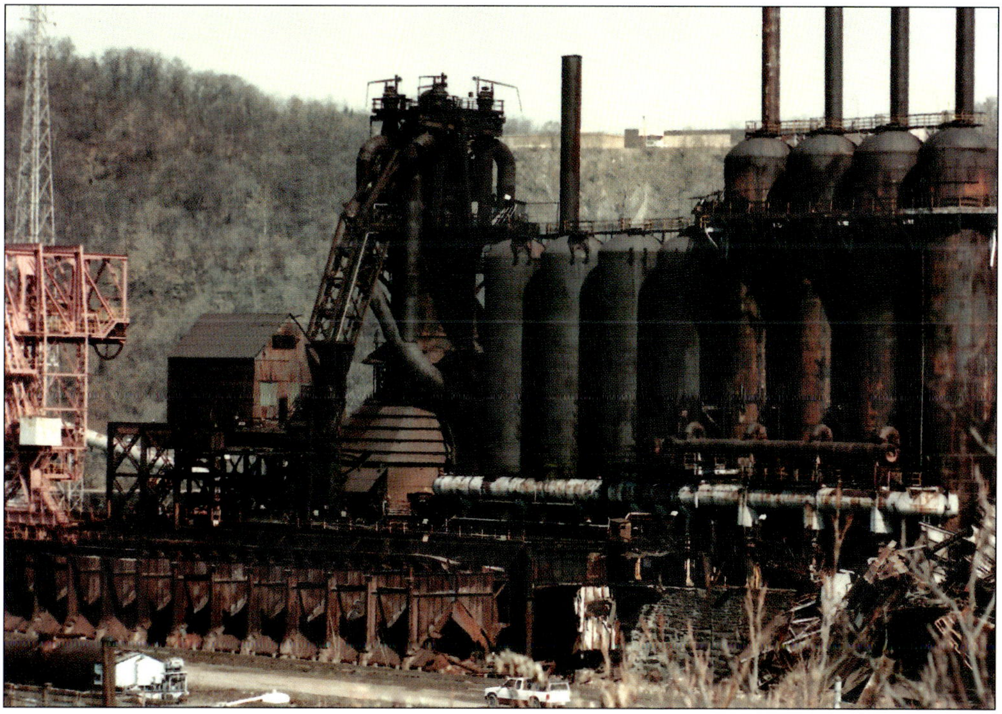

At Duquesne, Blast Furnace No. 1 used its four Cowper hot-blast stoves, along with the stoves' draft stacks, to preheat the air that is then blown into the furnace under pressure. At the bottom is the stocking trestle where iron ore, coke, and crushed limestone are weighed and sorted into skip hoist bins and delivered to the top of the furnace.

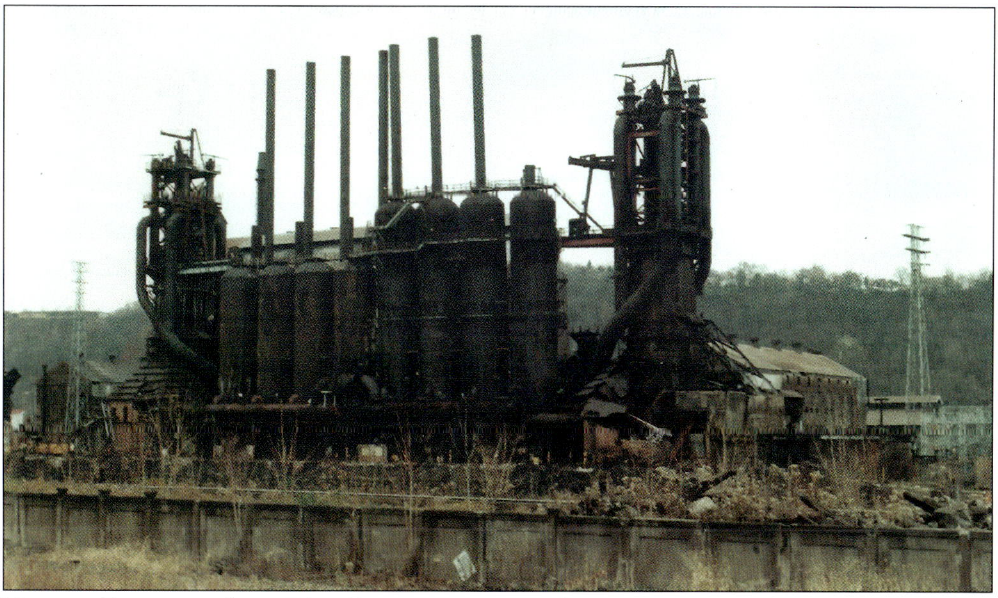

Here are Duquesne plant's Blast Furnaces Nos. 3 and 4. Note there are four Cowper hot-blast stoves instead of the usual three. At this point, the hoist houses have been removed from both furnaces. It looks like some of the truss from the skip hoist incline is hanging on the side of Blast Furnace No. 4. This photograph was taken from State Route 837 around 1990.

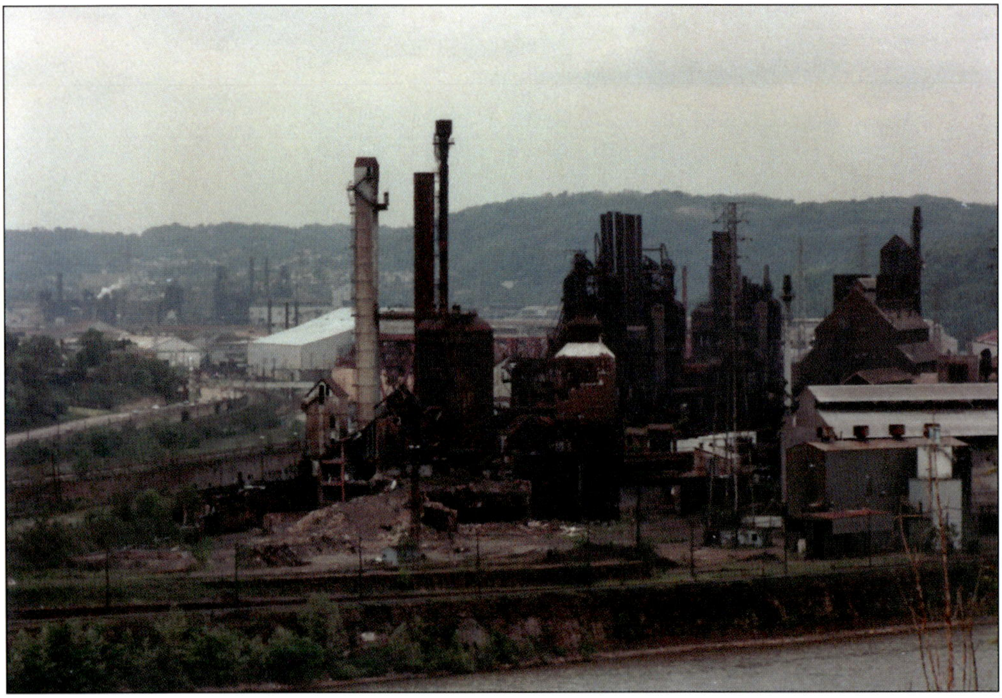

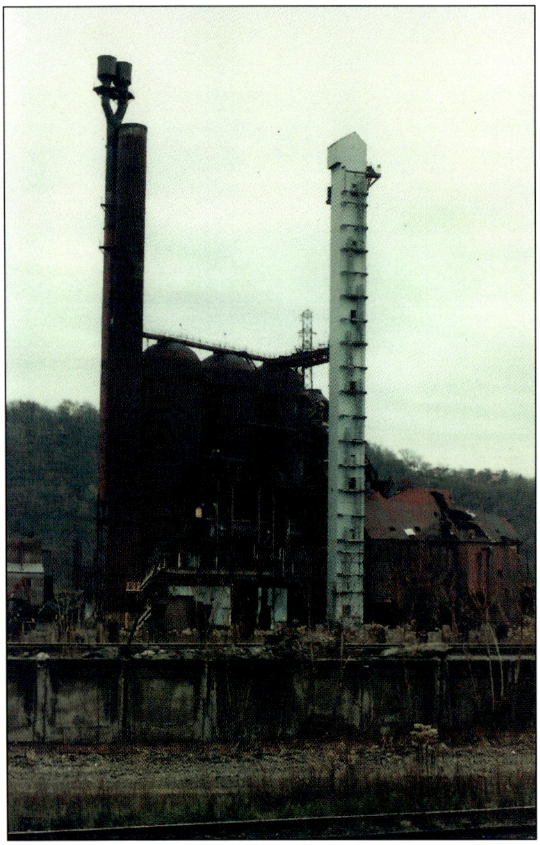

Pictured are the remains of Dorothy No. 6, the largest blast furnace; it was named after the wife of Leslie Worthington, a former president of US Steel. The furnace has been demolished. Only the elevator shaft, flare tower, draft stack, and Cowper stoves remain. After winning a company award in 1983, in 1984, it was economically unfeasible to operate it, and it was closed.

In this photograph, the elevator shaft, flare tower, draft stack, and Cowper hot-blast stoves are all that remain of furnace Dorothy No. 6. On December 29, 1983, US Steel announced it would shut down the Duquesne plant. In May, steel making was halted, eliminating 1,200 jobs. On October 19, 1984, a government- and labor-commissioned feasibility studies costing more than $220,000 was revealed. The first results were positive, and demolition was postponed until June 1985. The final financing report came on January 8, 1986. Costs would be $305 million to restart the plant, which was not feasible.

Three

Homestead Plant
The World's Biggest

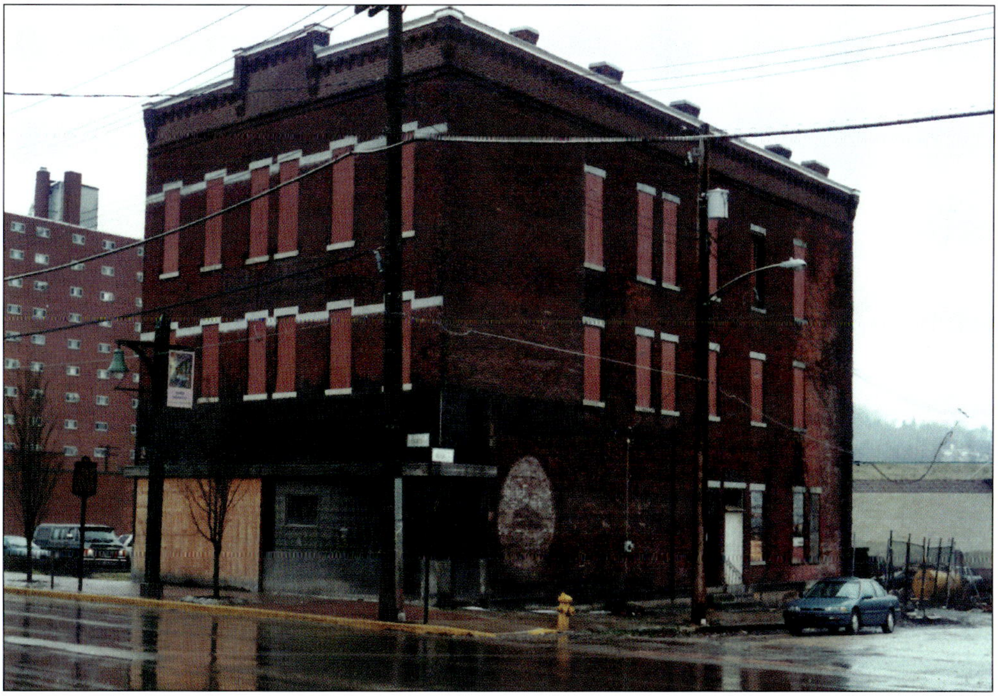

Pictured is the Columbia Hotel, opened in 1892 and located at 623 East Eighth Avenue and Heisel Street in Homestead, Pennsylvania. It was the headquarters for the Amalgamated Association of Iron and Steel workers during the lockout and the battle of Homestead. A telegraph line there was used by reporters to tell the world what was happening at the plant. Today, the building is used by Rivers of Steel Heritage Corp. for its offices.

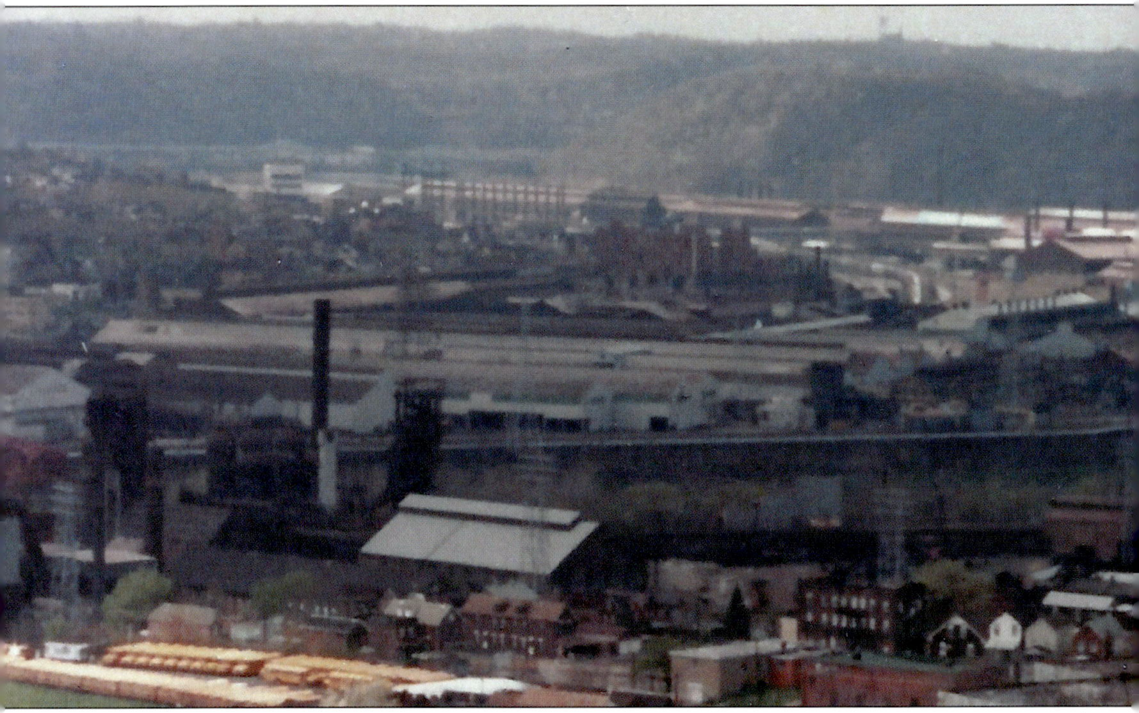

This 1984 panorama view of the Monongahela River valley was taken before the Homestead steel plant was demolished. Carrie Blast Furnaces Nos. 6 and 7 are seen on the left, and the

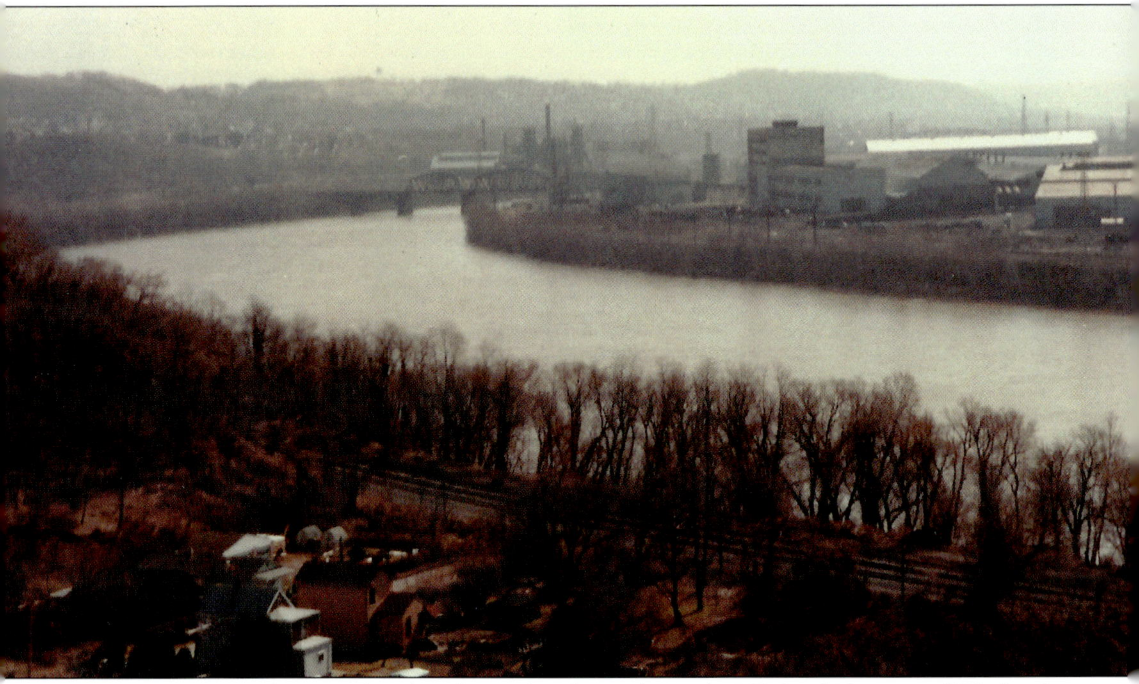

This panorama photograph, taken from "the Slags," shows the US Steel Homestead Works along the Monongahela River. Duck Hollow is seen on the near side where Nine Mile Run empties into the river by the railroad trestle. Open-Hearth Building No. 5 is seen on the right, and Carrie

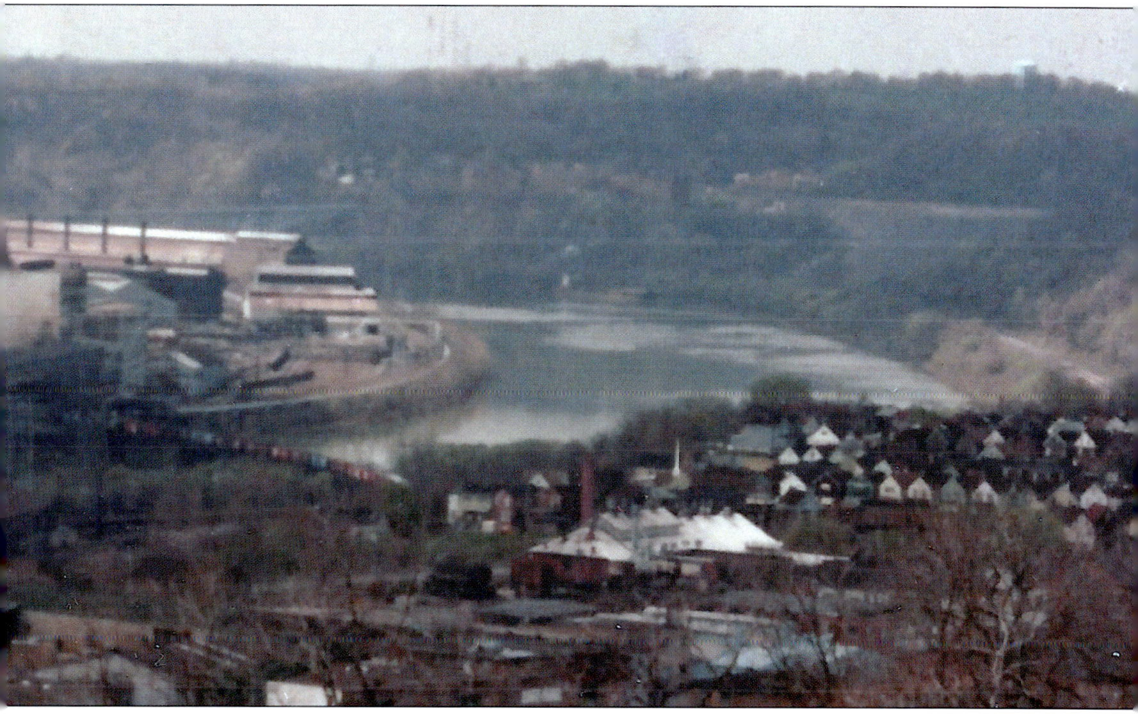

rest of the Homestead plant is across the river. A train can be seen going over the Pinkerton's Landing Bridge.

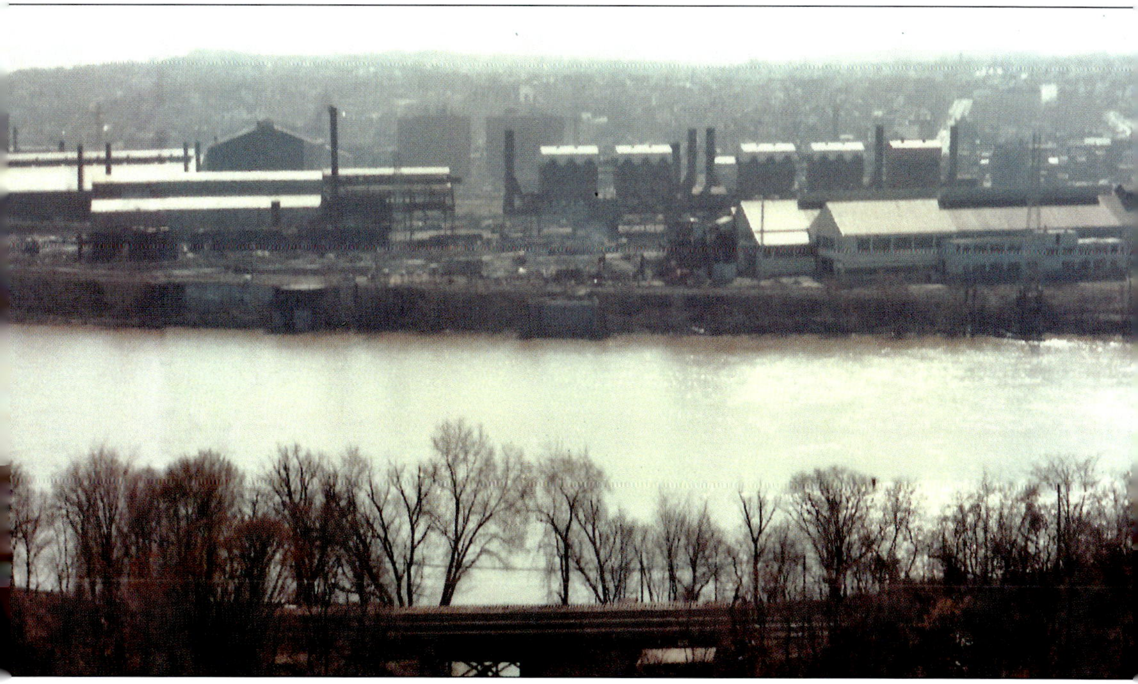

Furnace is in the distance, just around the bend on the left. The plant continues on the right, past the Homestead Grays Bridge.

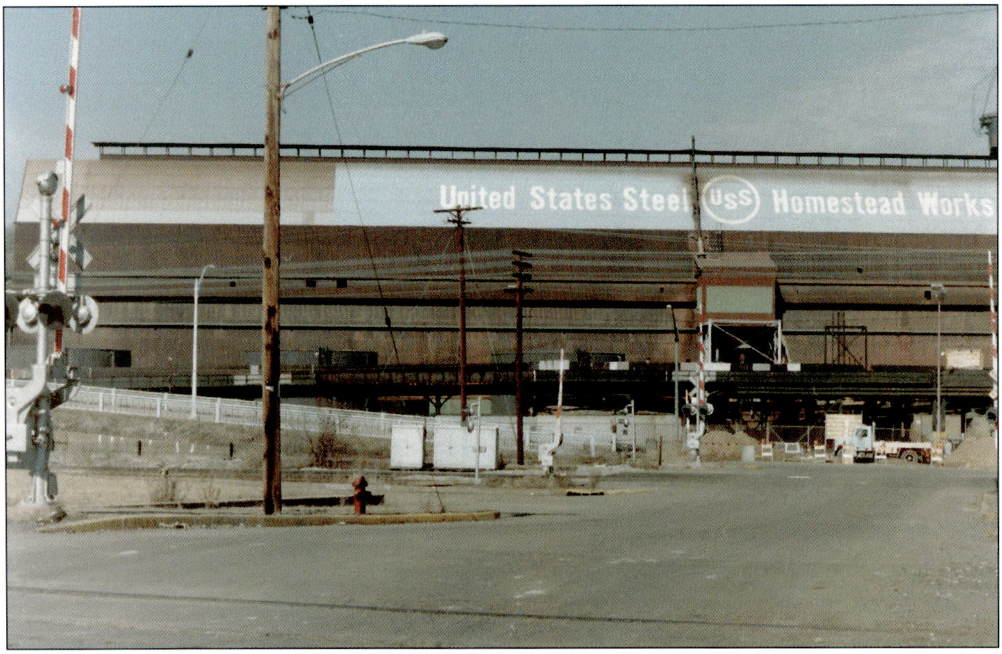

This photograph shows the exterior of the Open-Hearth Building No. 5 at the Amity Street Gate by the Homestead Grays Bridge. One can see "United States Steel Homestead Works" has been painted in white on the roof. The Amity Street Gate was a main employee entrance and exit point for the plant. Many strike pickets were held there.

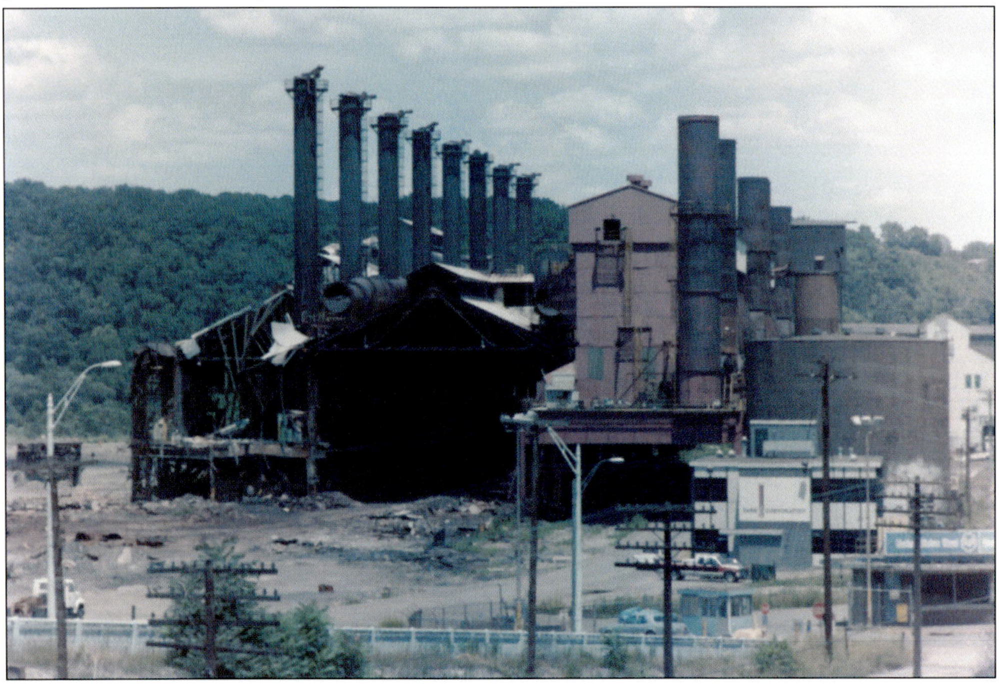

Here, Open-Hearth Building No. 5 is being demolished. This c. 1989 photograph was taken from the Homestead Grays Bridge, looking east. The front of the building has been taken off. This is at the Amity Street Gate. Note the security booth.

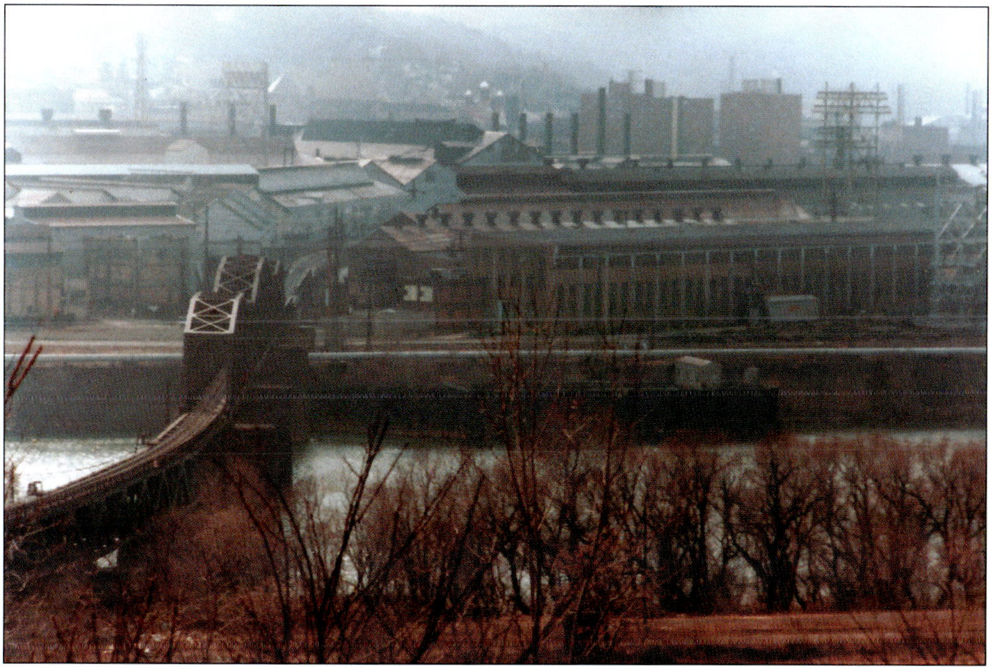

Seen across the Monongahela River from Swissvale, Pennsylvania, is the US Steel Homestead Works. Also pictured is the Pittsburgh & Lake Erie Railroad (P&LE) Bridge at Munhall (CSX, Pittsburgh Subdivision); it is a Parker Pratt through-truss bridge with Pratt deck trusses. Its vertical clearance is 51.6 feet. The bridge was built around 1883 and upgraded pre-1902.

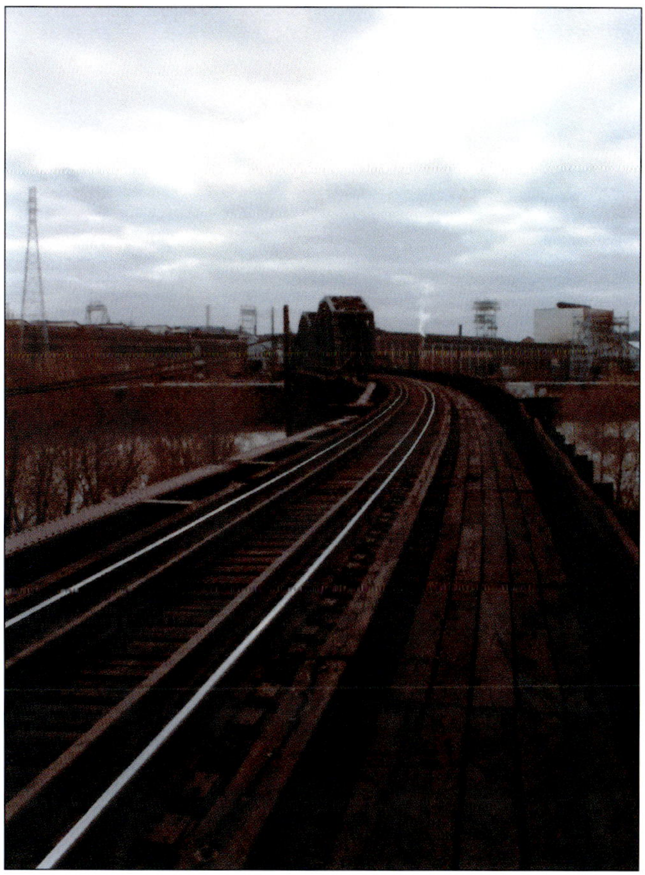

Pictured is the P&LE Railroad Bridge at Munhall; this bridge is also known as the Pinkerton's Landing Bridge. Some people would use the wooden side passageway on the Swissvale side to gain entrance to the Homestead plant in order to loot scrap metal (copper cables) after it closed on July 25, 1986.

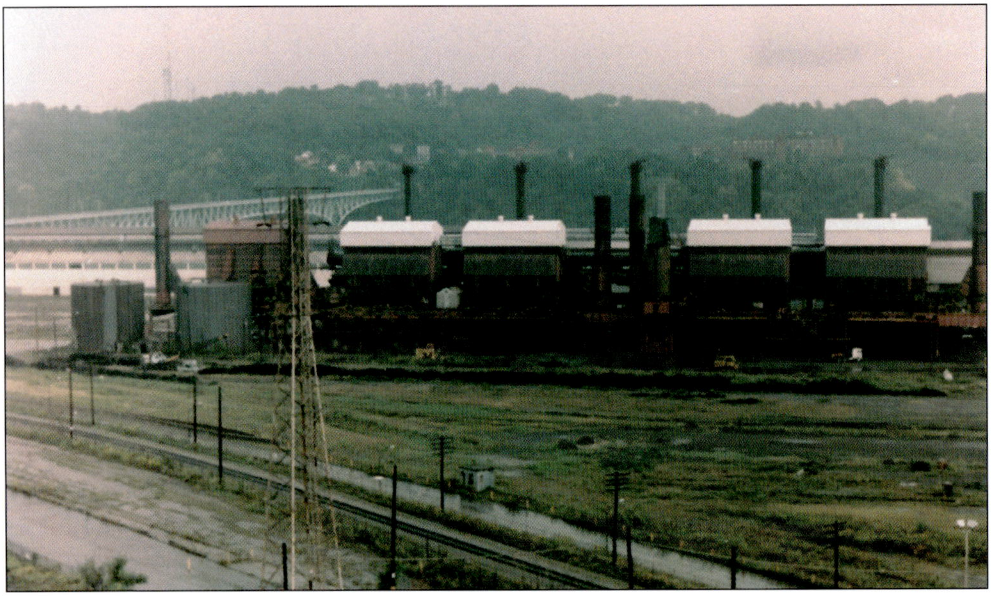

In this photograph of Open-Hearth Building No. 5 at the Homestead plant, one can see the electrostatic precipitators on the side of the building.

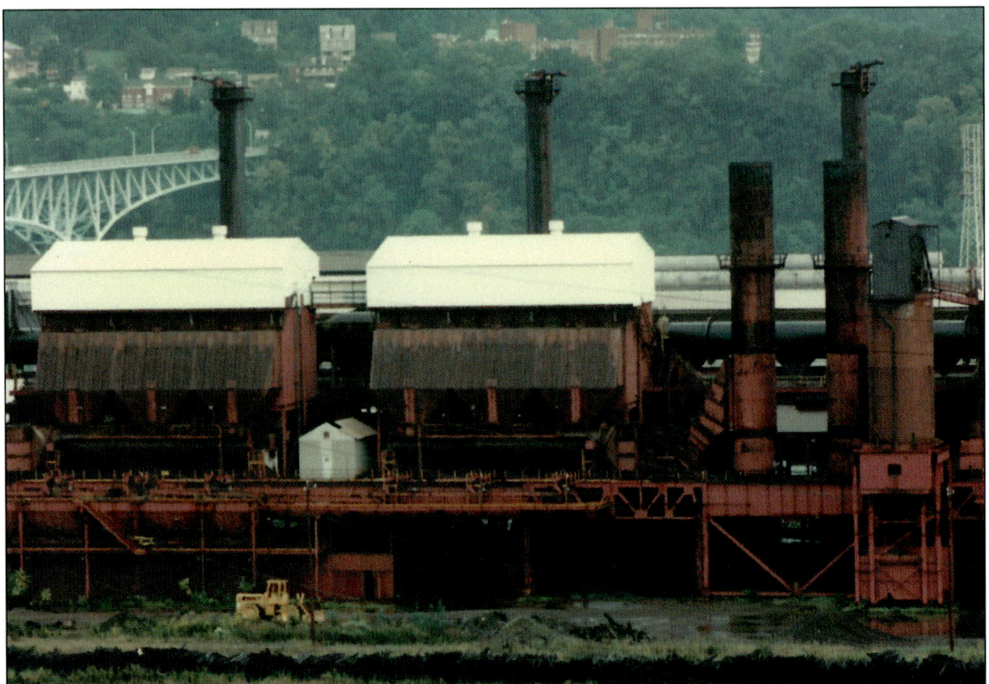

Seen around 1989, Open-Hearth Building No. 5 was painted red. Note the discarded railroad ties in the foreground and the Homestead Grays bridge on the left.

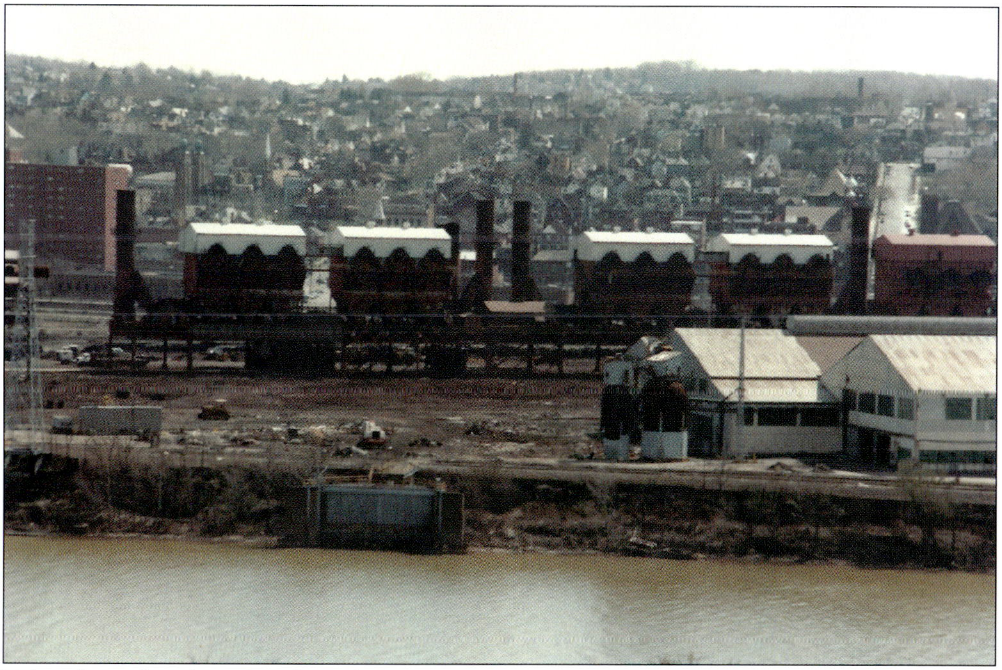

Taken from the Slags, found on the north bank of the Monongahela River, this photograph is of the Homestead plant. Note that many of the buildings have already been demolished. Slag is a generic term for the waste by-product of the steel-making process. After demolishing the Homestead mills and clearing the site, this today is known as the Water Front Mall

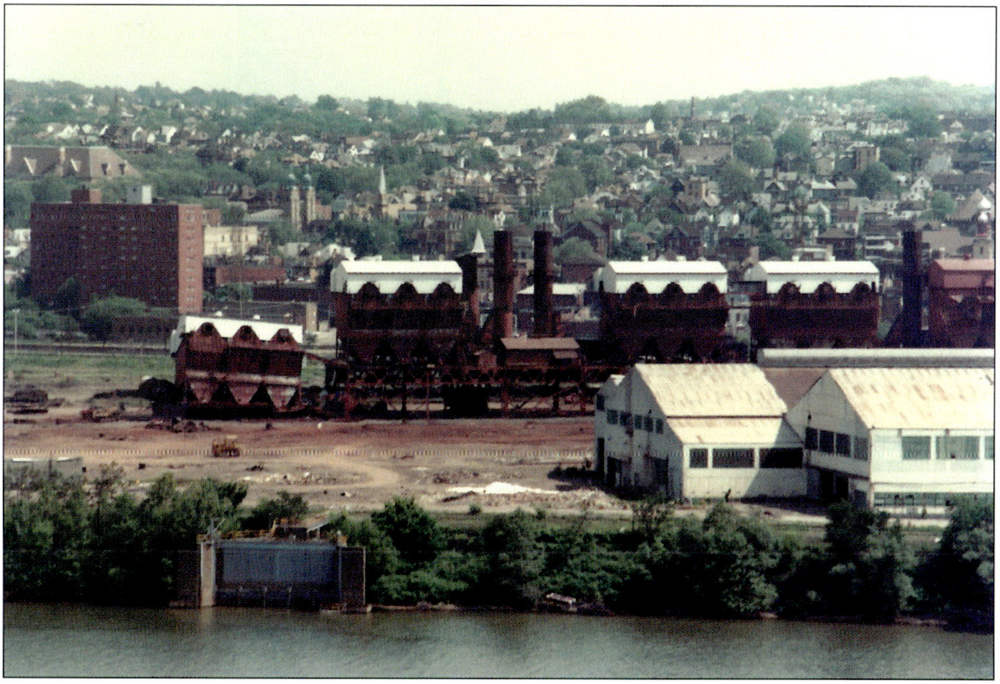

Here is a section of the Open-Hearth Building No. 5 that has been demolished. Little by little, the buildings were taken down, broken up, scooped into trucks, driven away, and gone forever.

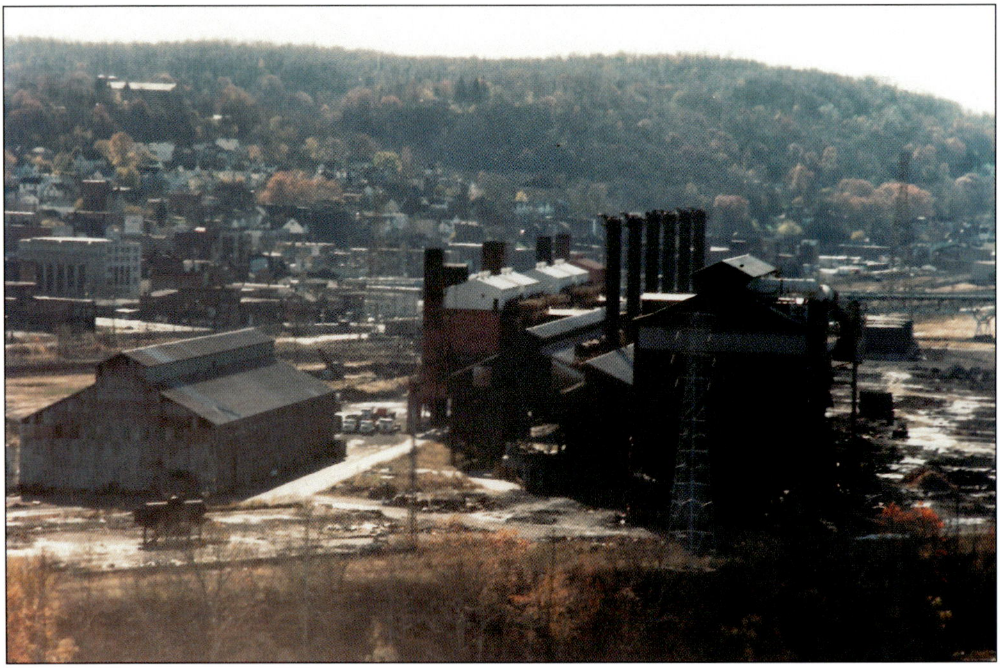

This c. 1990 view from across the Monongahela River is of the US Steel Homestead Works as it is being demolished. The structures are, from left to right, the boiler shop extension, Open-Hearth Building No. 5, and the plate-treating building closest to the river.

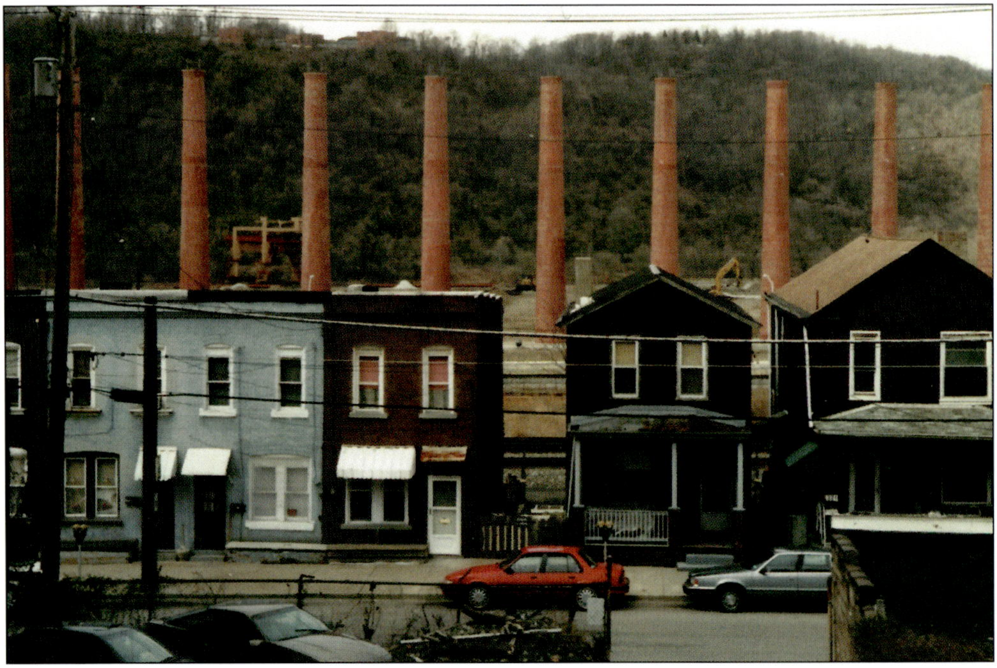

These 12 venting stacks belong to the soaking-pit furnace building at the Homestead plant. This photograph, taken in West Homestead from Eighth Avenue, looking down across West Seventh Avenue, shows some homes in front of the plant. The stacks are among the few things left standing after the plant was cleared. They were built by the M.W. Kellogg Co. of New York.

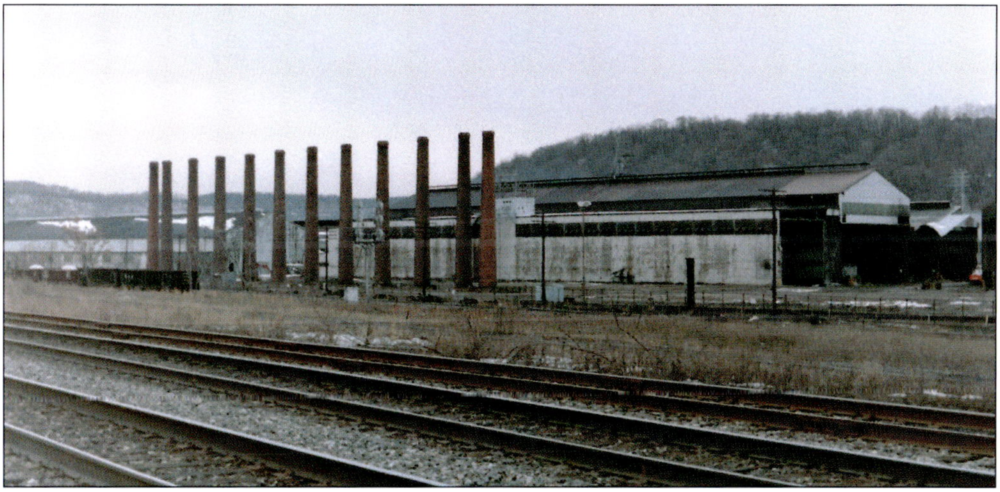

With few other surrounding buildings still intact, the 12 venting stacks are depicted here around 1989. Note the old-style railroad "Christmas-tree" signal seen near the center of the photograph. Located in West Homestead, the three sets of train tracks are now running Norfolk Southern trains.

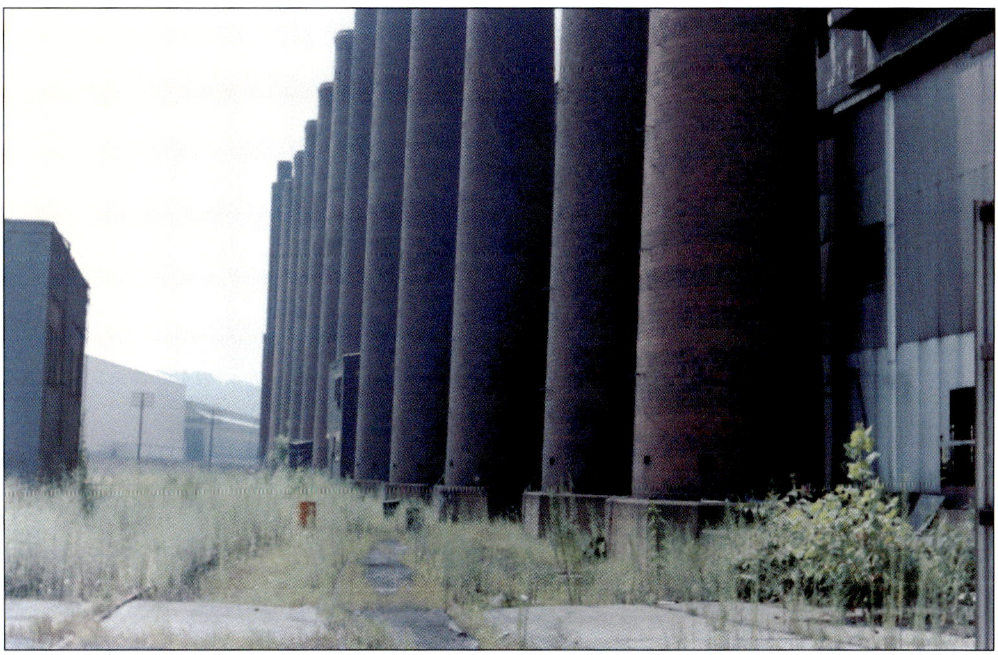

These venting stacks are all that remain today of the 45-inch soaking-pit furnace building. Forty-five inches is the width of the steel slabs coming off the universal plate mill. The author took this photograph while working for Process Instruments, salvaging temperature chart recorders, which were refurbished and sold to other users.

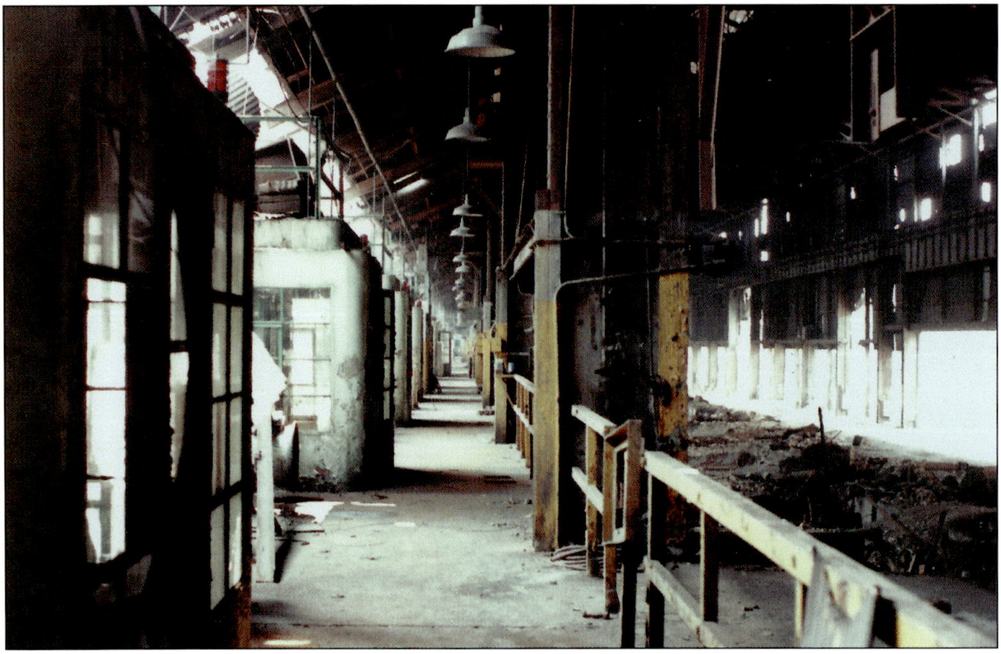

Here is the east end of the 45-inch soaking-pit furnace building. This building was 100 feet wide by 600 feet long and contained 20 recuperative natural and coke oven gas-fired pit furnaces lined up in two horizontal rows. The control rooms are seen on the left; everything on the right side has been removed. Today, everything is gone.

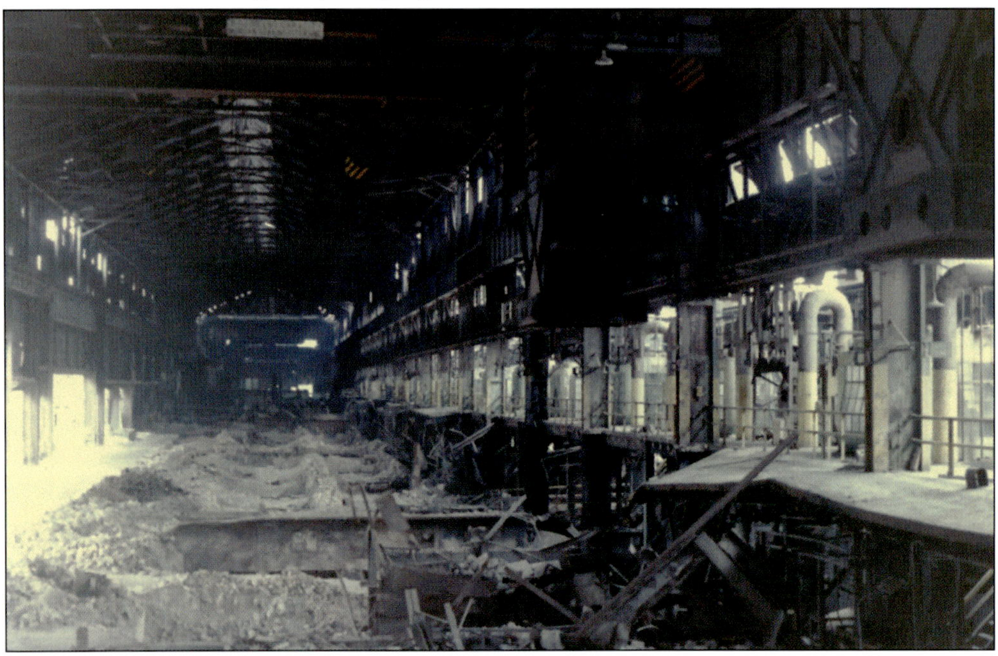

Looking west inside the 45-inch soaking-pit building. Up to six ingots were charged by a 25-ton overhead crane into each furnace. The furnaces were 15 feet wide by 16 feet long by 13 feet deep. They were designed to equalize the temperature of the ingot. After heating for up to 18 hours, the ingots were removed from the furnaces and taken away in an ingot buggy.

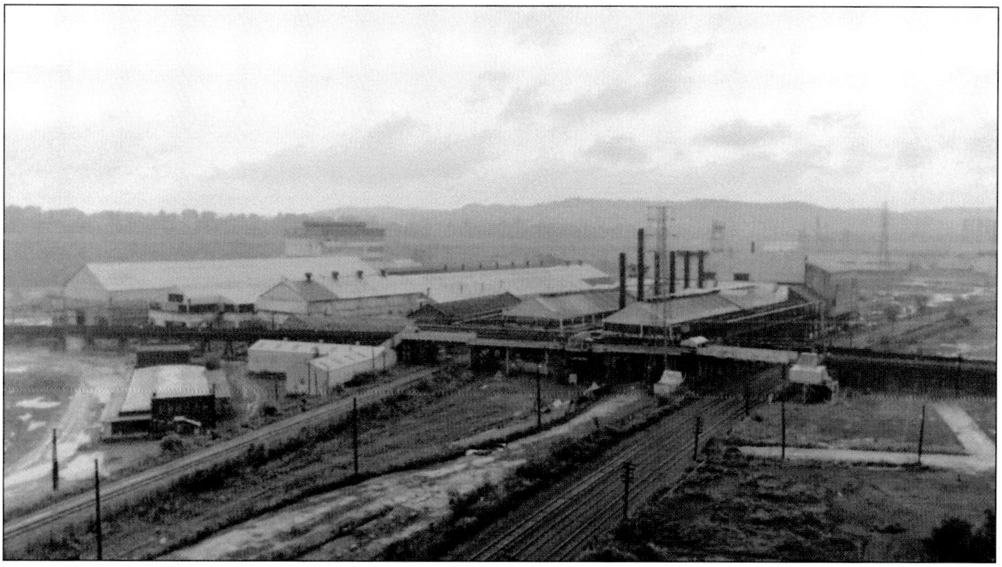

Taken from the roof of an Allegheny County Public Housing Building, this photograph is of the US Steel Homestead Works. Behind the trestle are Machine Shop No. 2, the vertical furnace building, and pit furnaces. Carrie Furnace site is seen in the distance on the right on this overcast day.

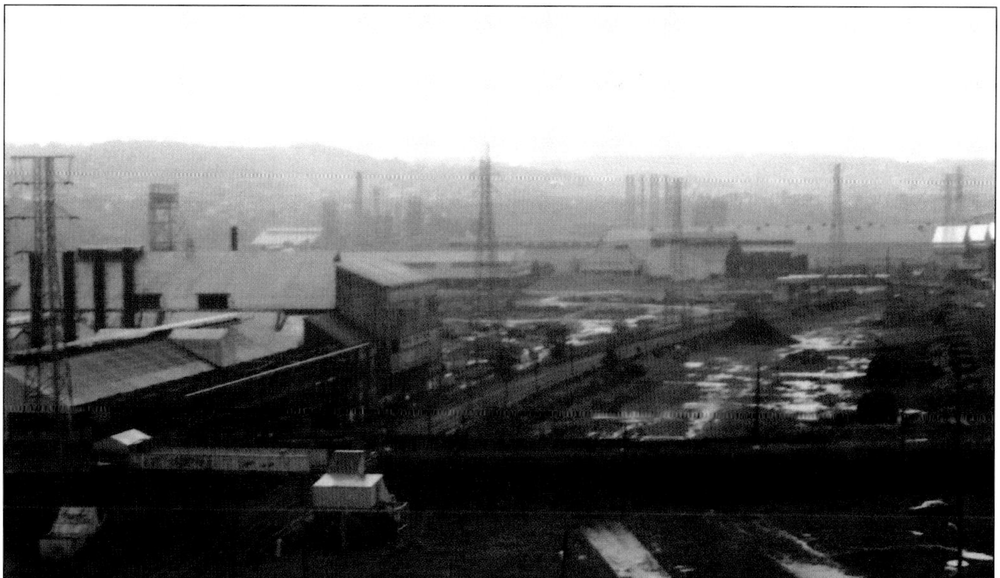

Again, this view is looking northeast from the roof of the Allegheny County Public Housing Building, which stands along Dixon Street in Homestead, one block west of the Rivers of Steel Heritage Building.

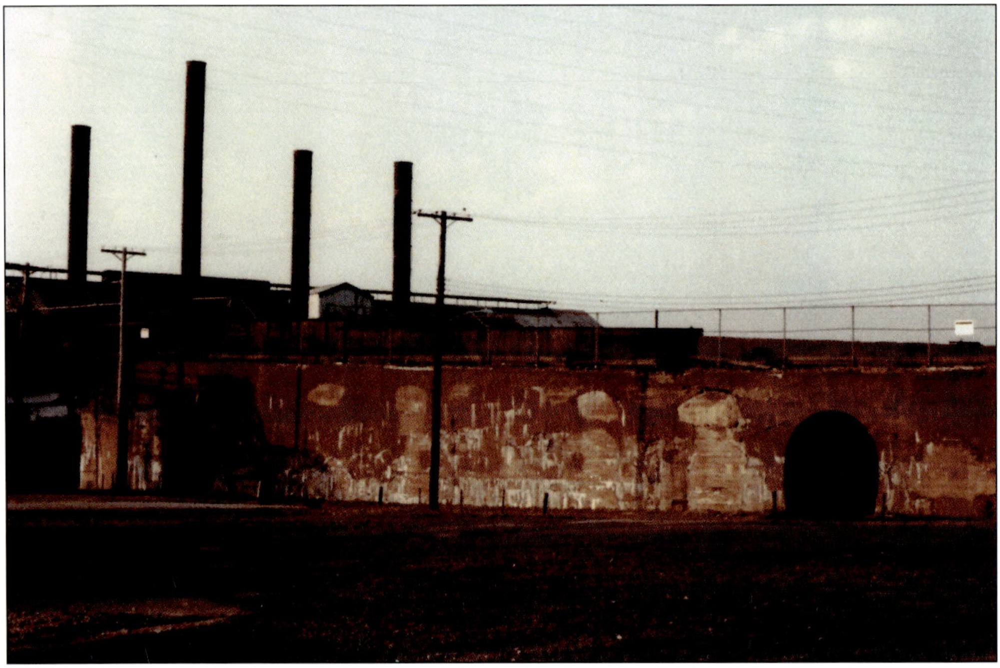

At the Homestead plant, the train trestle of Open-Hearth Building No. 4 forced the need for the famous Hole in the Wall Gate on City Farm Lane; the archway on the right gave access to the plant and the paymaster's office. The gate on the left is named Wagon Gate.

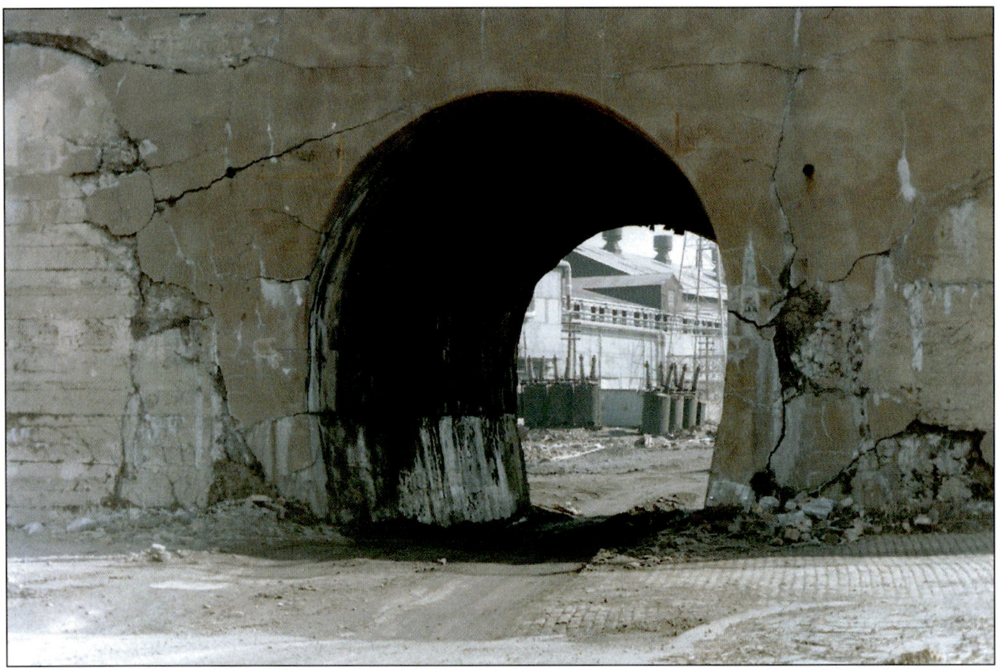

This archway below the train trestle of Open-Hearth Building No. 4 let trucks through to the plant to be loaded with scrap metal from the Homestead plant. The street in front of the wall is City Farm Lane. The archway was known as Hole in the Wall Gate. Another gate over to the left and closer to the tracks alongside the plant was called the Wagon Gate.

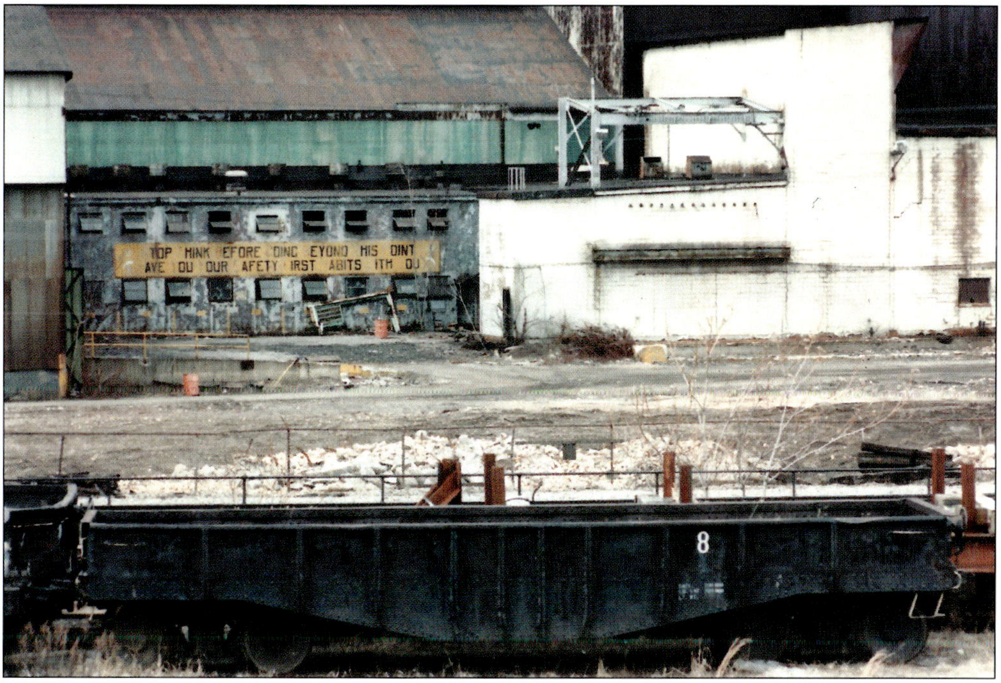

Some prankster removed the first letter of each word in the safety slogan painted on the side of this building. It should read as follows: "Stop think before going beyond this point / have you your safety habits with you."

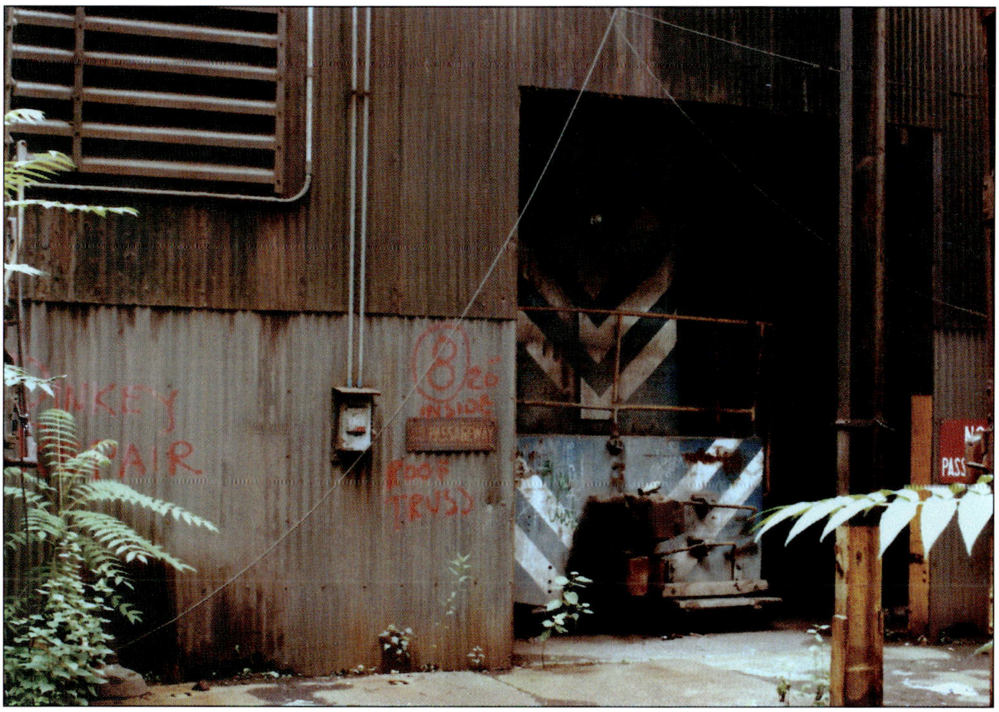

Seen in May 1989, a train engine sits in a "loco" repair shop waiting for its fate to be decided while signs warn that there is "no passageway."

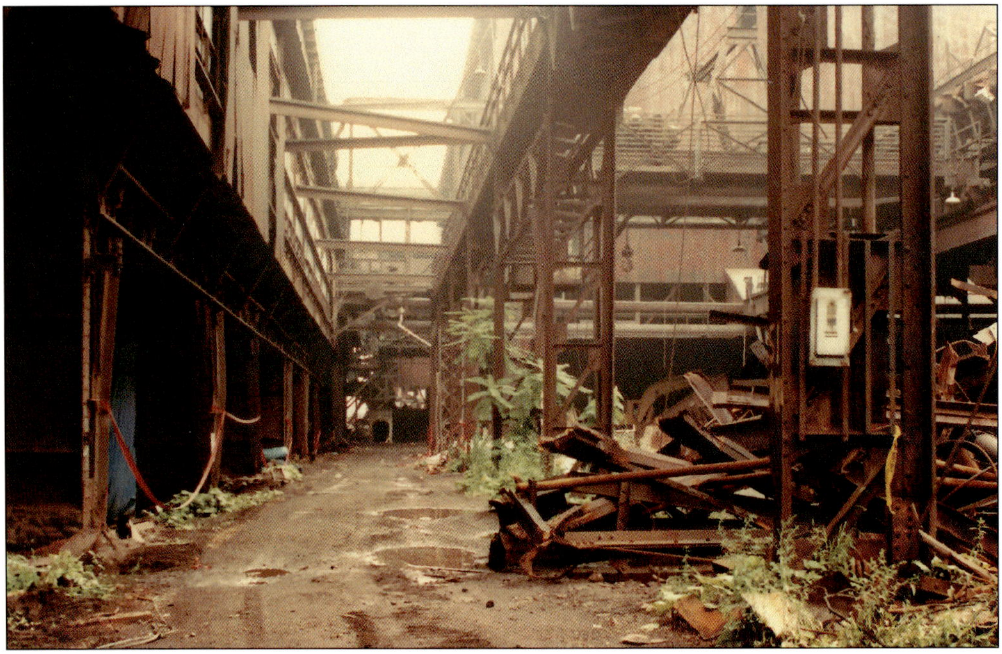

This photograph was shot in June 1989, but it is unclear where inside the Homestead plant it was taken. The plant was deserted except for the workers demolishing it. They did not want anyone taking photographs that might show safety violations or other causes for legal action to be brought against them, delaying the demolition of the property.

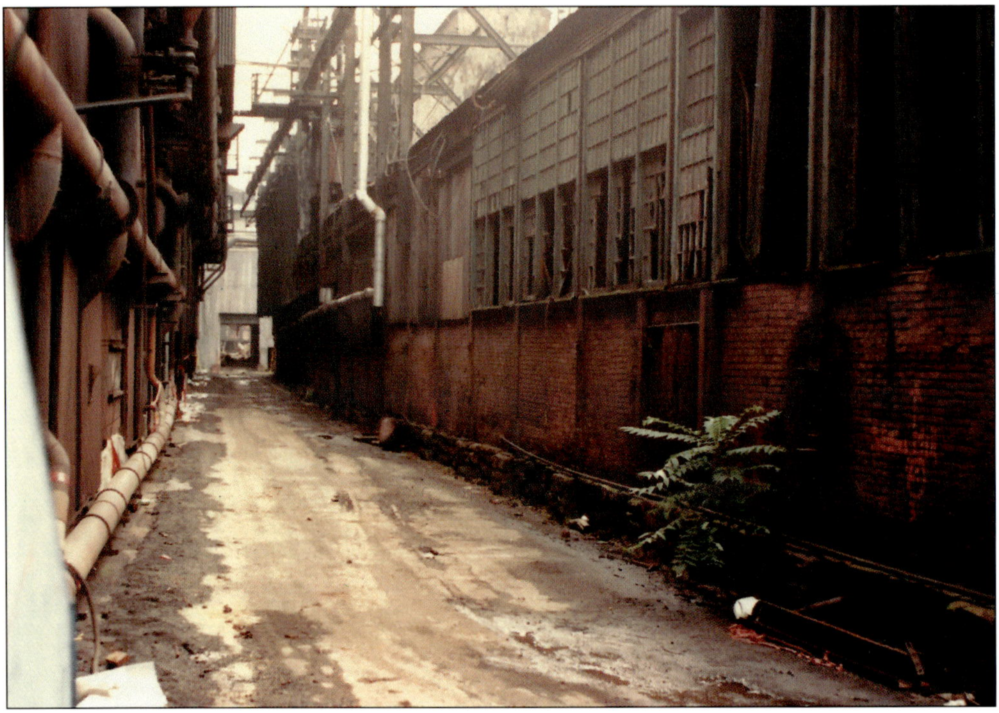

This June 1989 photograph was taken shortly before the US Steel Homestead Works was all torn down and trucked away.

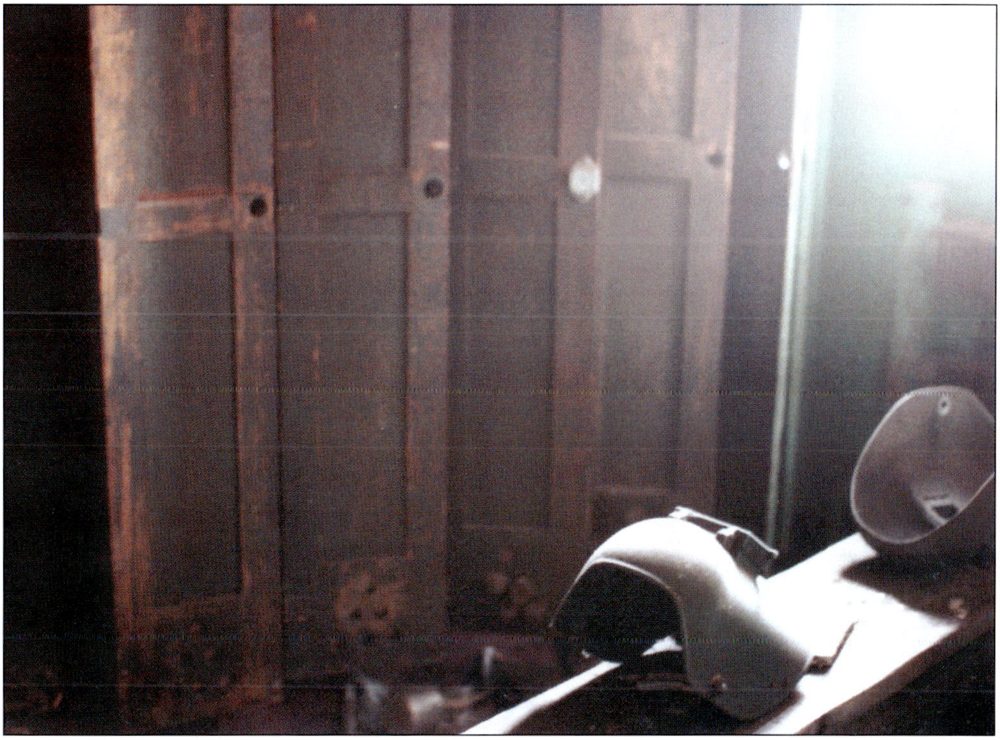

Pictured is a workers' locker room. Here, men showered to clean up before going home for the night (or day). Note the welding masks left lying around. The plant was so big that it was in several different municipalities, so where a person's locker was determined what city one paid taxes to.

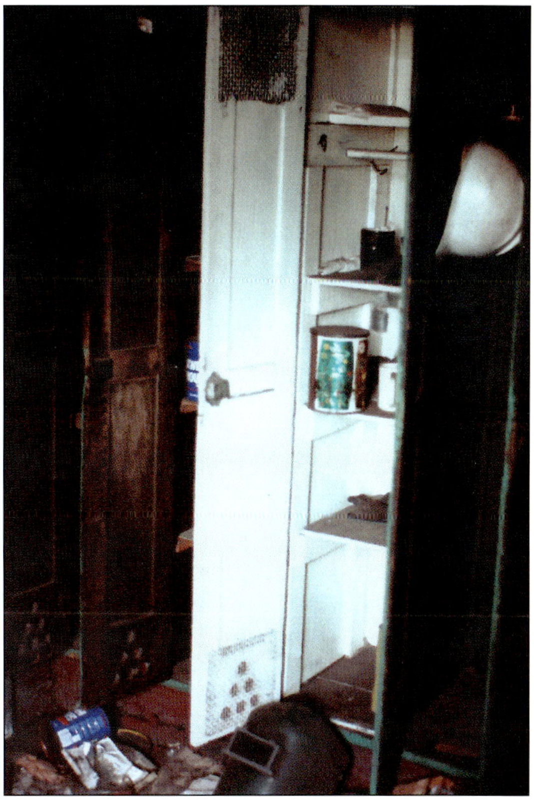

Occasionally, people referred to lockers as "steelworkers' shanties." When plants were closing, it was not uncommon for whole departments to shut down without notice; on the day they were let go, workers were informed to "clean out your lockers, remember to return the locks." Some people were fired after 24 years and some months, meaning "There's no 25-year pension for [them]." Those men and women were told good luck getting a new job at age 55.

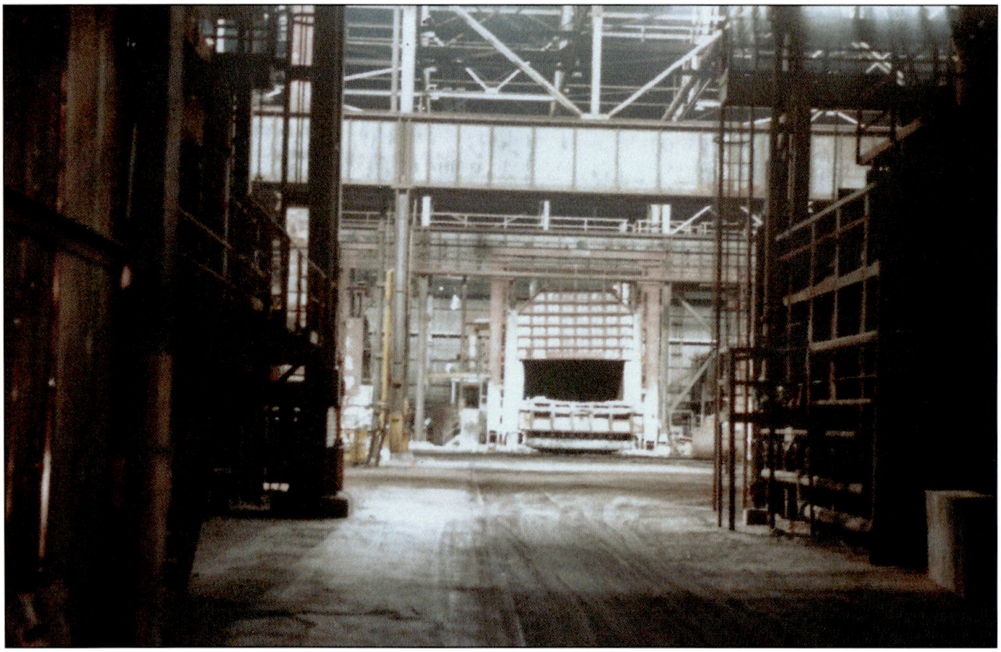

Here is Forge Shop Building No. 2, originally constructed in 1941. One can see a reheater furnace straight ahead. The building was drastically different from what it had been five years before when it was in operation. It closed in 1984. The noise, heat, sweat, and passion that were found in this building are only memories now.

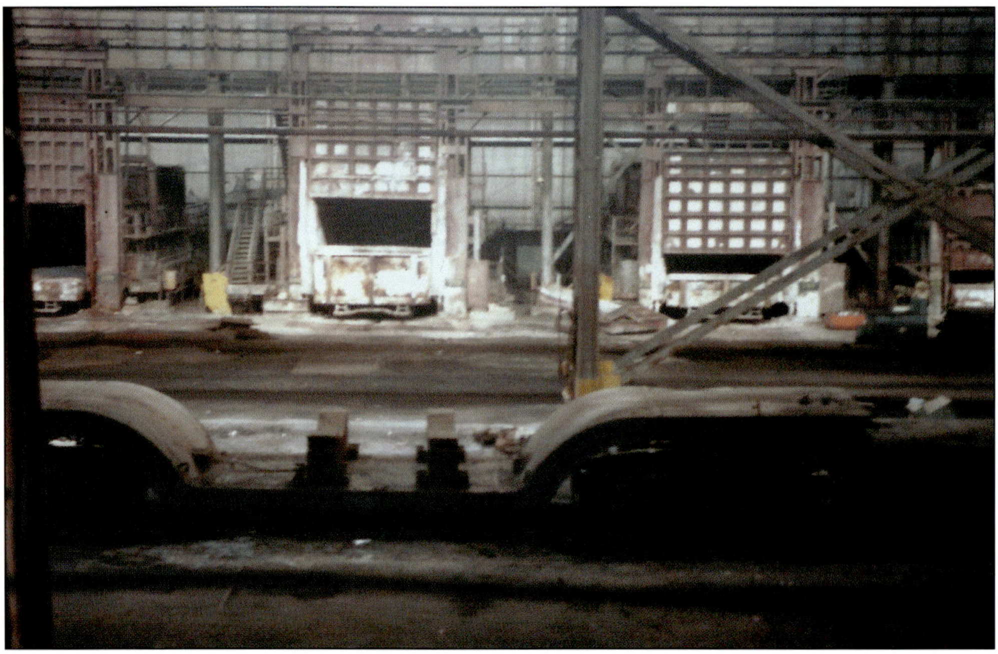

In 1989, a railcar sits in the middle aisle of Forge Shop Building No. 2 at the Homestead plant. It was used to transport ingots of hot steel to the slabbing mill. It must have been quite a sight to see hot ingots, weighing several tons, being moved by crane from the furnaces onto railcars moving along the center aisle.

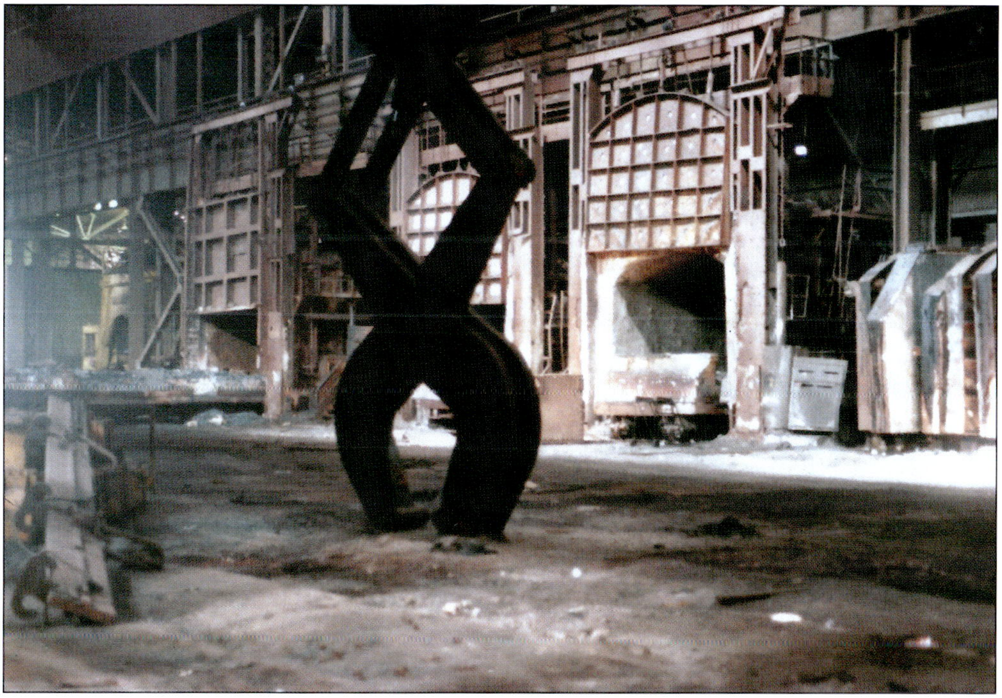

These are giant-size tongs. Found in Forge Shop Building No. 2, the tongs stood seven feet, seven inches tall. They were possibly used for placing metal ingots from the furnaces onto railcars that were moving through the shop to the slabbing mills.

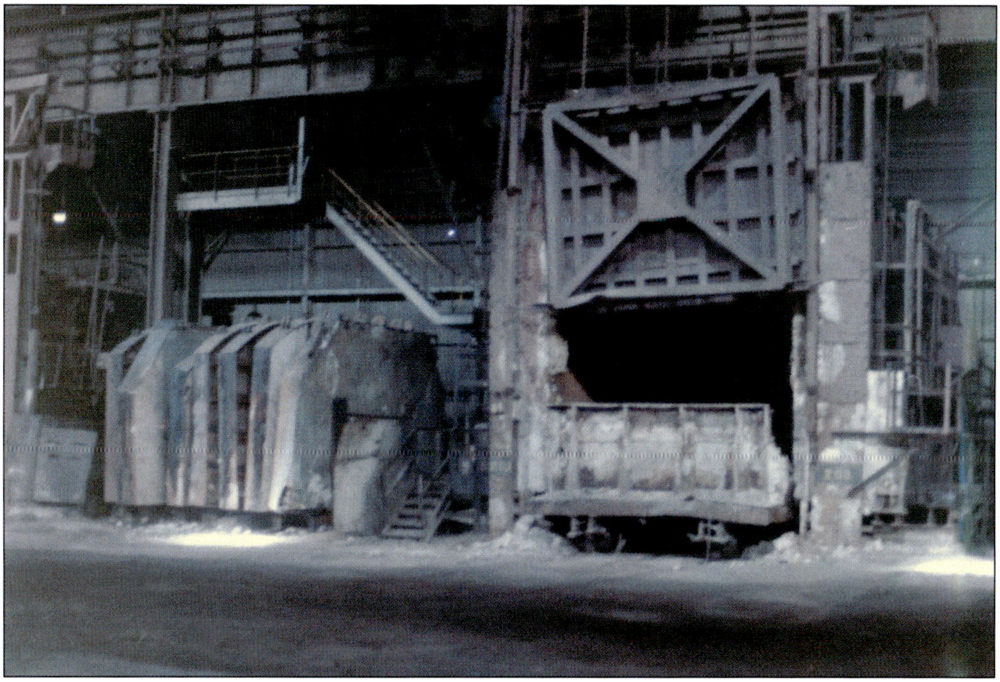

Here is another photograph of a reheater furnace in Forge Shop Building No. 2 at the Homestead plant. Sometimes these type of furnaces are referred to as "car-bottom" furnaces.

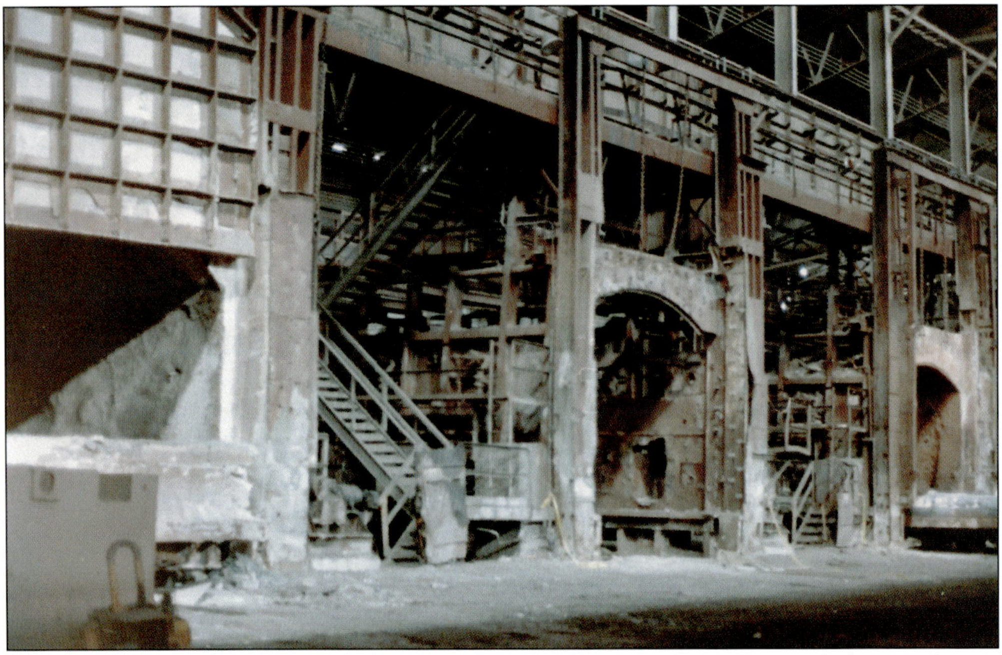

This is a 1989 interior view of Forge Shop Building No. 2 at the Homestead plant. Reheater furnaces are lined up along the north wall of the building. Quiet and cold five years after shutting down in 1984, this building would soon be demolished and trucked away as scrap.

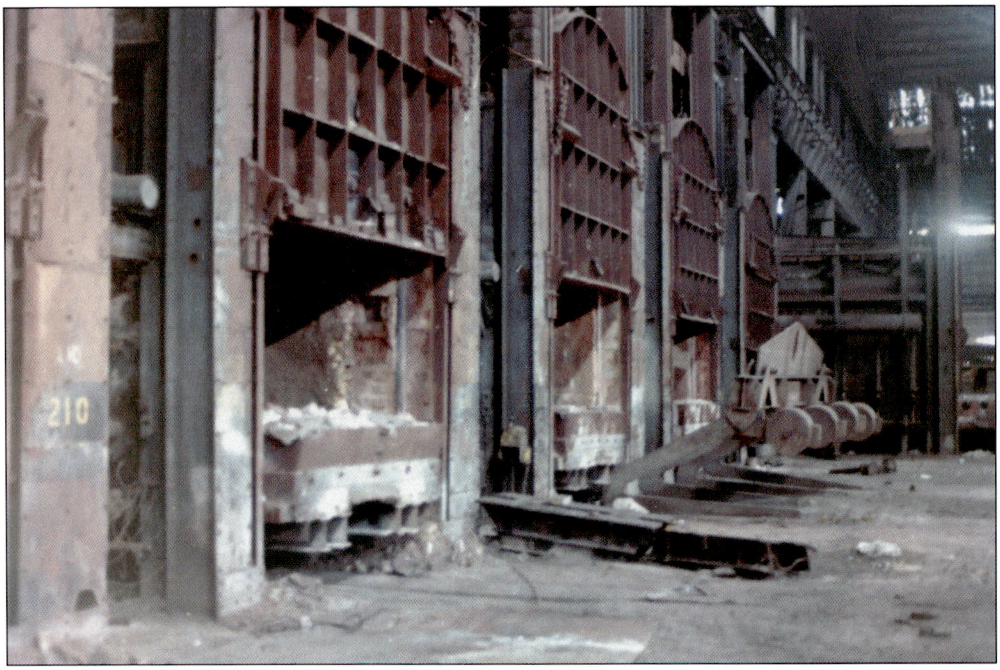

This picture was taken a little farther down the line of reheater furnaces in Forge Shop Building No. 2. Note "210" painted in yellow on the left side of the photograph; what the number 210 stood for, no one can remember.

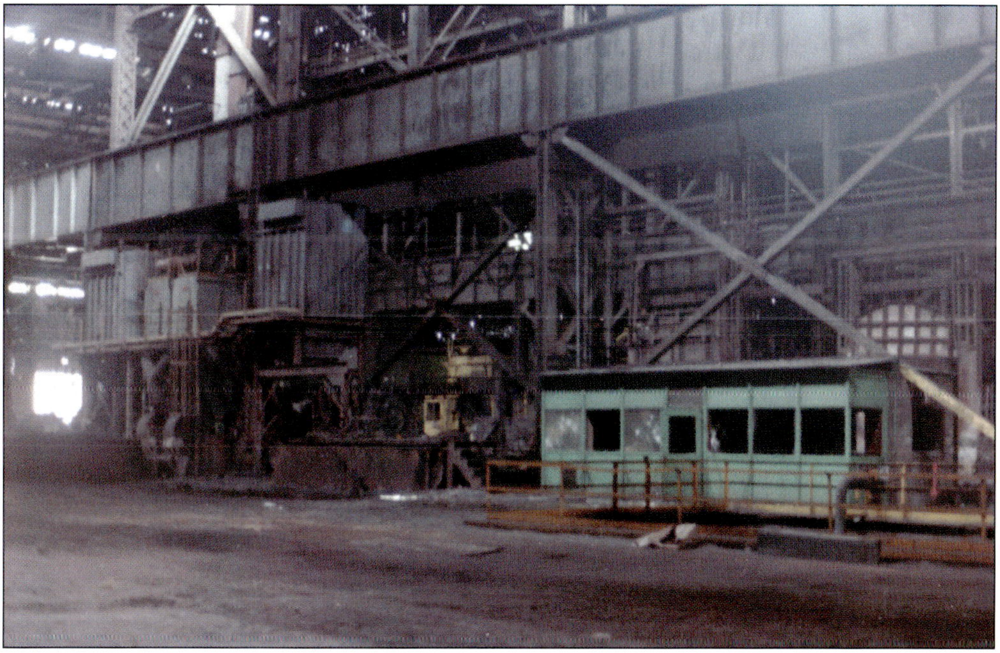

Still farther east inside Forge Shop Building No. 2, one can see a green-painted control room in the center aisle. Another reheater furnace is located behind it on the north side of the building.

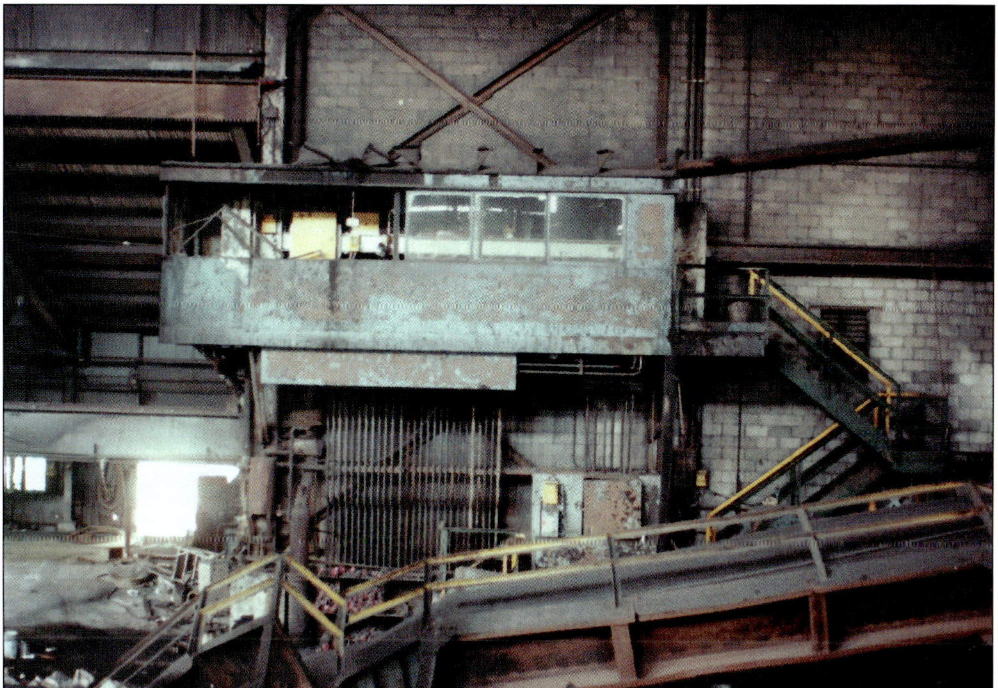

In this photograph, one can see a green-painted control room mounted on the side wall of the building. It was awe inspiring to walk along this gigantic building where so much industrial activity had taken place. Once a loud, hot, and bright place, it has been forever quieted, chilled, and darkened.

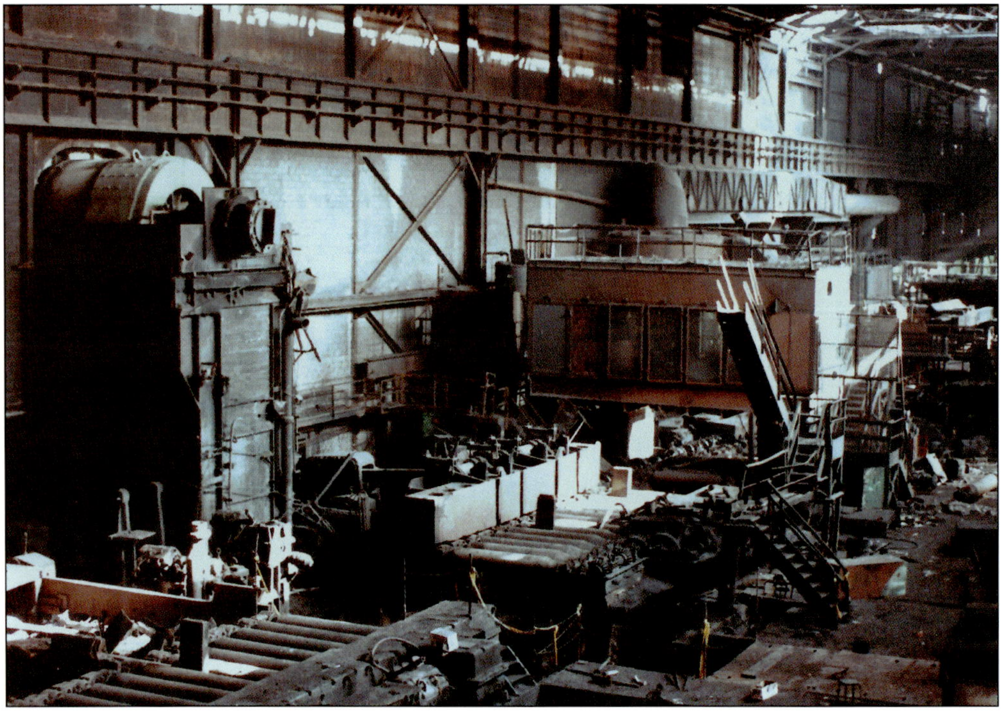

This is the inside of the 45-inch slabbing mill building, where ingots were worked into slabs.

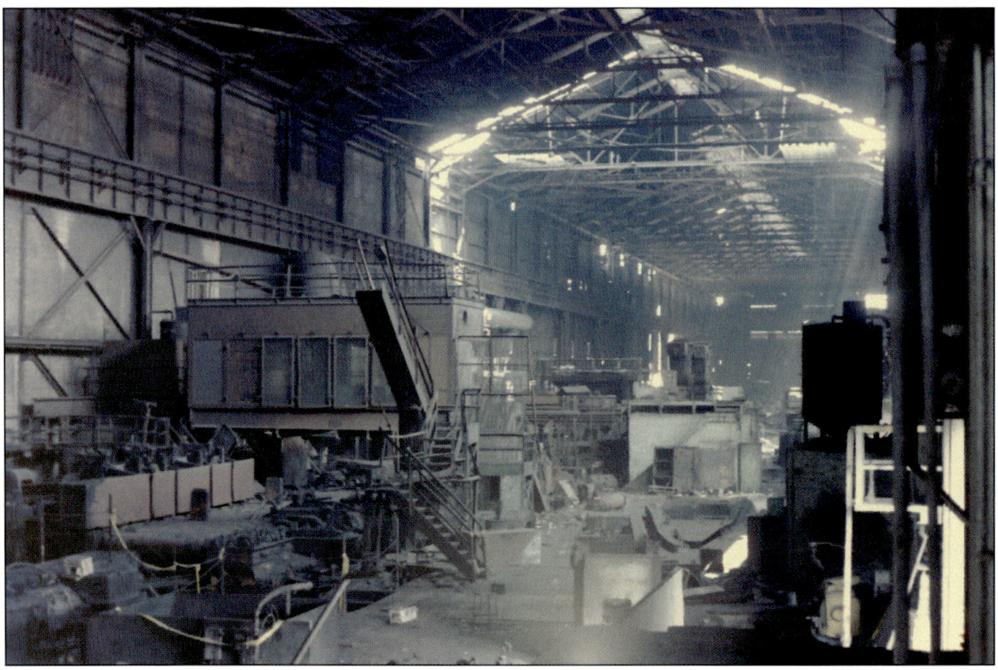

Finished slabs were stacked, and buggies took them to the other side of the wall to the 100-inch or 160-inch plate mill storage areas.

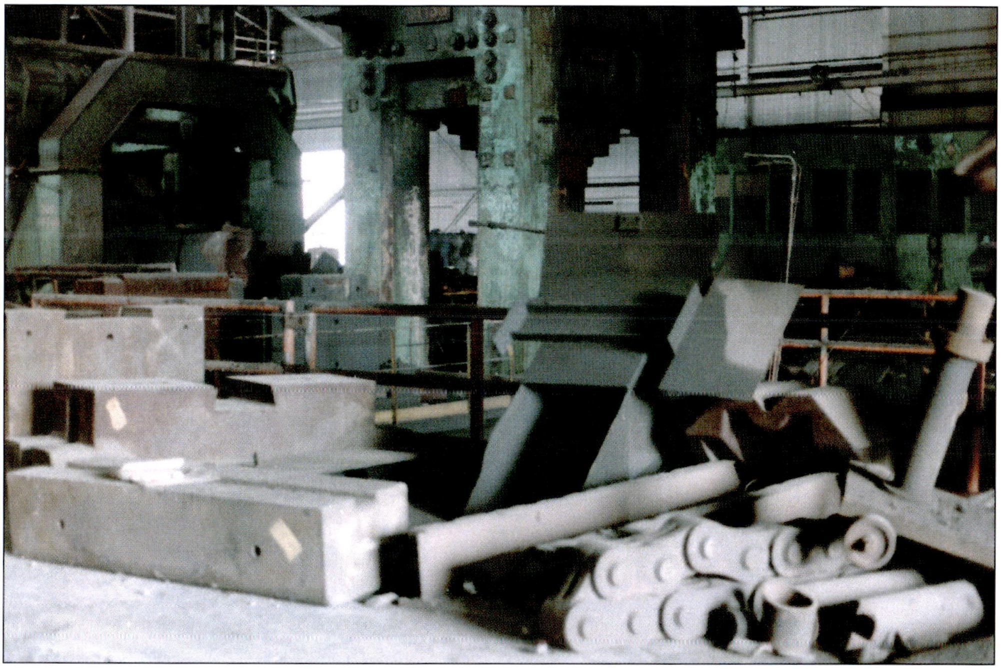

Pictured is the 4,000-ton press near the end of Forge Shop Building No. 2 and Press Shop No. 2. This machine is marked "Bliss," and next to it is the Alliance 200-ton manipulator for the 4,000-ton press. Note the roller chain, located at the bottom right of the photograph. It would have been wrapped around an ingot and used to lift it.

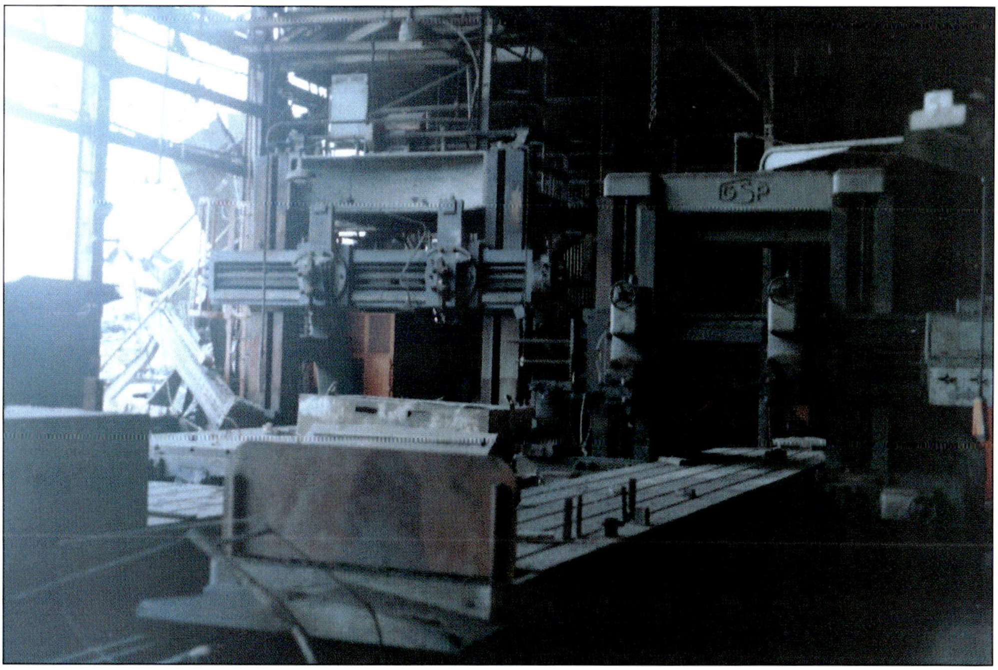

Another pair of machines photographed in the same area are these two metal planners. This brings an end to Forge Shop No. 2 and Press Shop No.2 at the Homestead plant.

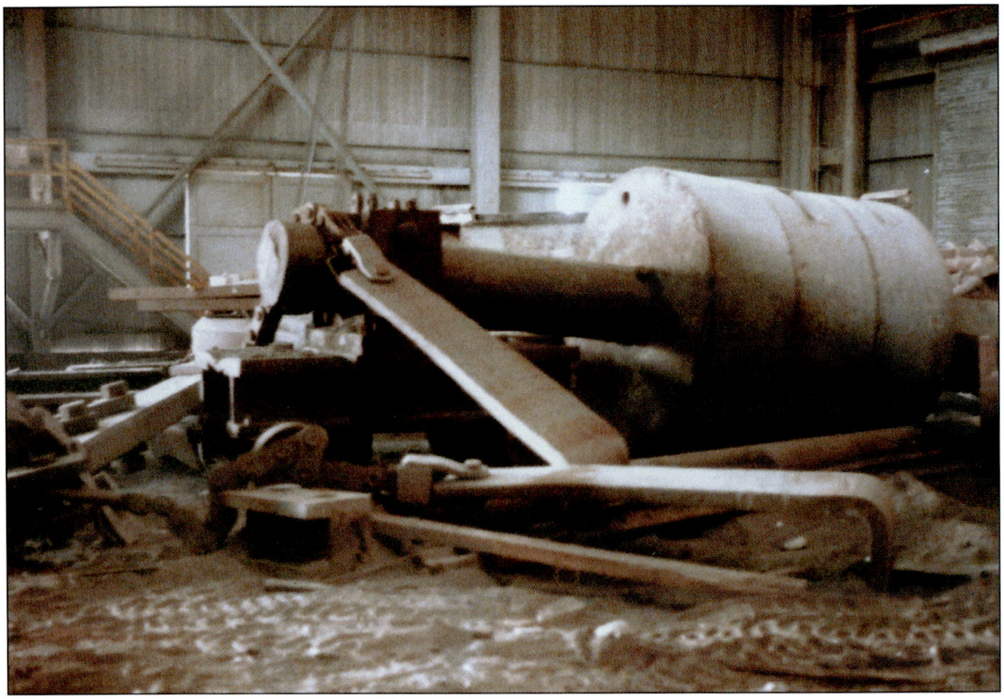

This huge counterweight was found near one end of the Central Maintenance Building lying on the ground with a roller chain wrapped around one end.

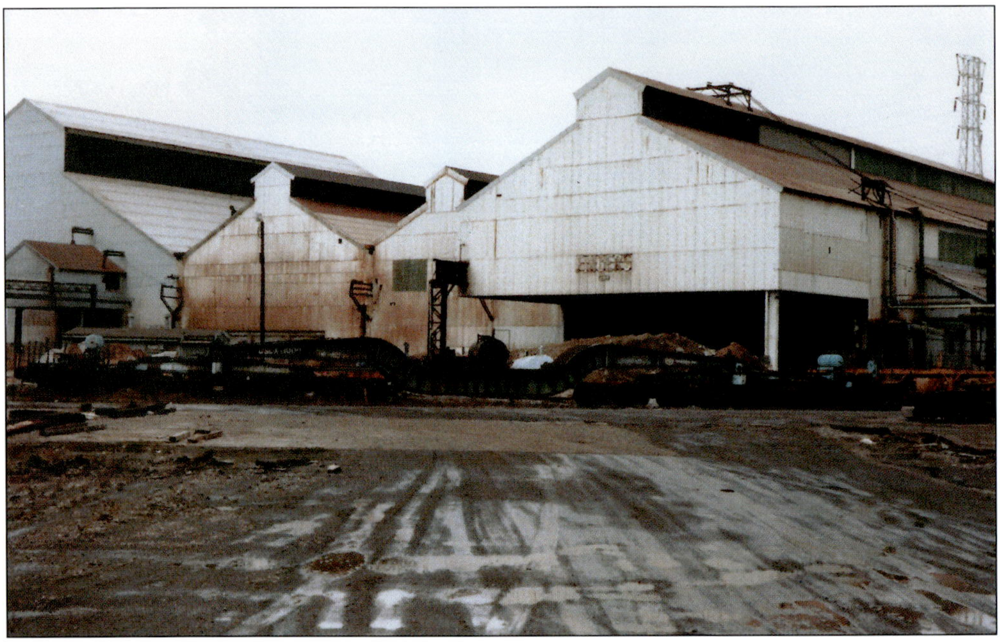

Here is an exterior view of the Central Maintenance Building, formerly the 30-inch mill, on the right. The railcar sitting in front of the building was made by the Pennsylvania Engineering Corp. in New Castle, Pennsylvania, in April 1971. It was built for extremely heavy loads such as forging ingots and molds weighing close to 500 tons.

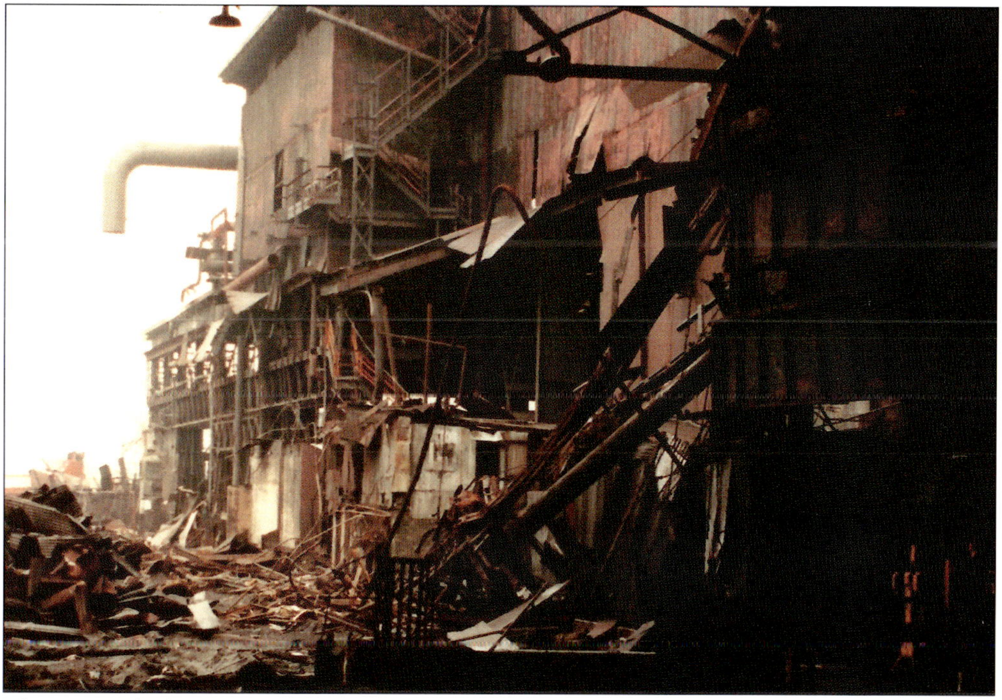

This view is looking out at the demolished side of a building at the US Steel Homestead Works. The famous Homestead steel plant, which, at one time, had been the largest steel plant in the world, made steel to construct the buildings, bridges, and ships that made this the greatest nation in the world. That gave Pittsburgh the world-renowned title of "Steel City." In 1989, the plant was sold for scrap.

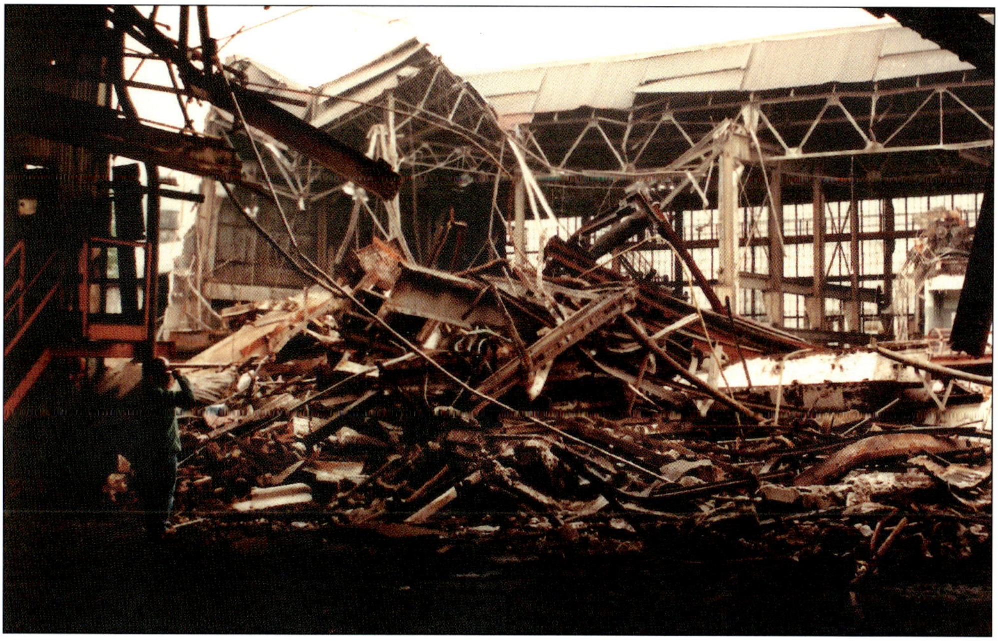

Here is a last look at a pile of rubble that used to be the mighty US Steel Homestead Works. Note in the lower-left corner the hardhat-wearing man saluting what is left of the steel plant.

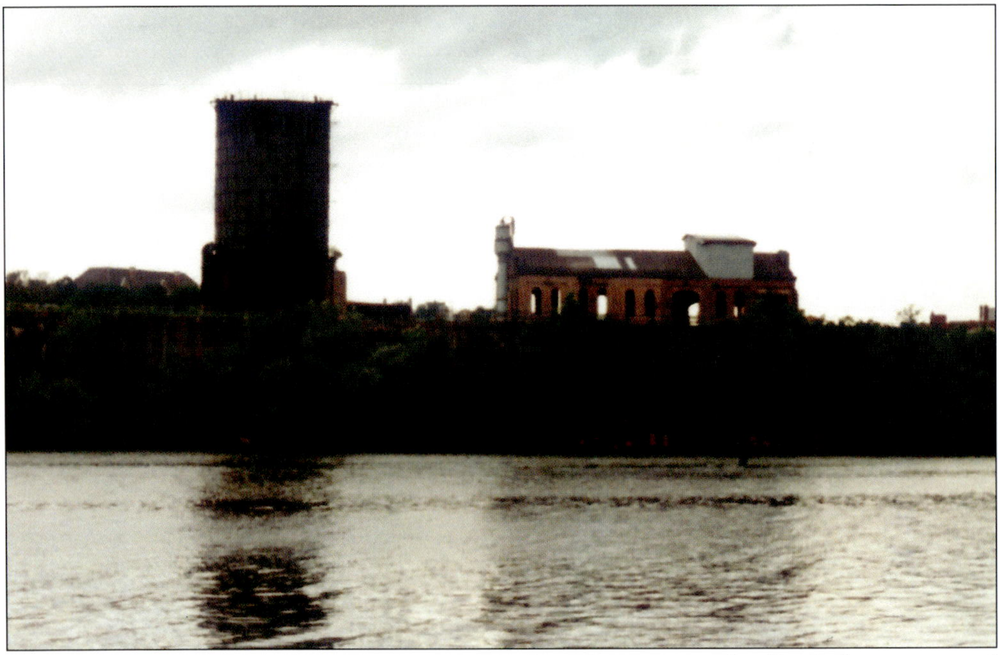

Pump House No. 1 is the only remaining structure from the Homestead mill, which once covered three miles of riverfront property. During World War II, it employed between 15,000 and 20,000 workers. The pump house is the site of the 1892 battle of Homestead where workers and townspeople repulsed a force of Pinkerton detectives who landed from barges on the river.

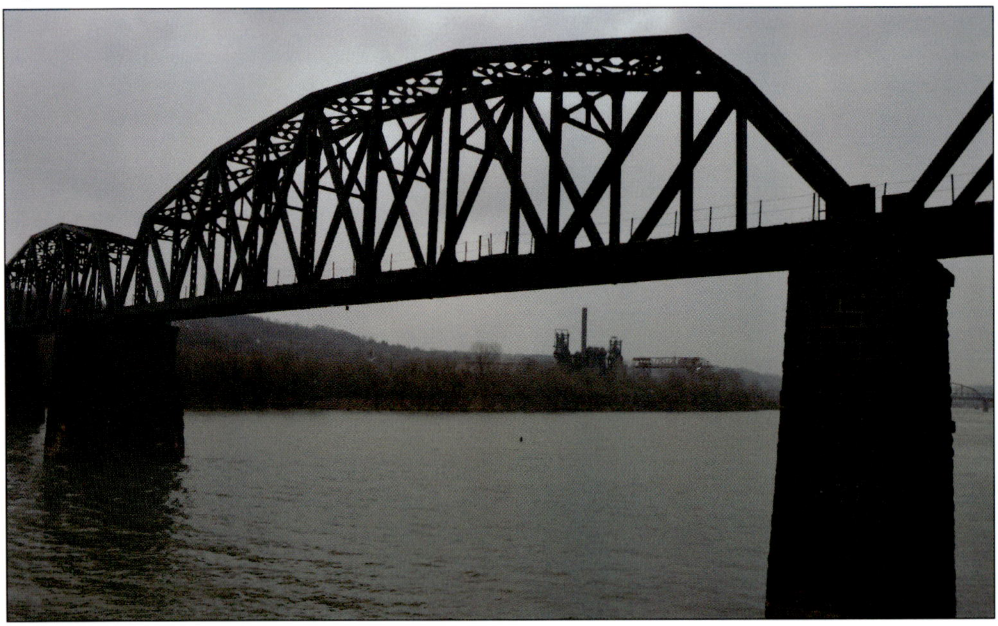

The Pittsburgh & Lake Erie Railroad Bridge at Munhall, also known as the Pinkerton's Landing Bridge, now carries the CSX Railroad, Pittsburgh Subdivision. The bridge crosses the Monongahela River between Munhall (south bank) and Rankin (north bank). Constructed with three deck trusses, there are two Pratt through-truss side spans and one Pratt through truss. The main span is 250 feet. Vertical clearance is 51.6 feet, and it was built in 1883 and was upgraded pre-1902

Four

Carrie Furnace National Historic Landmark

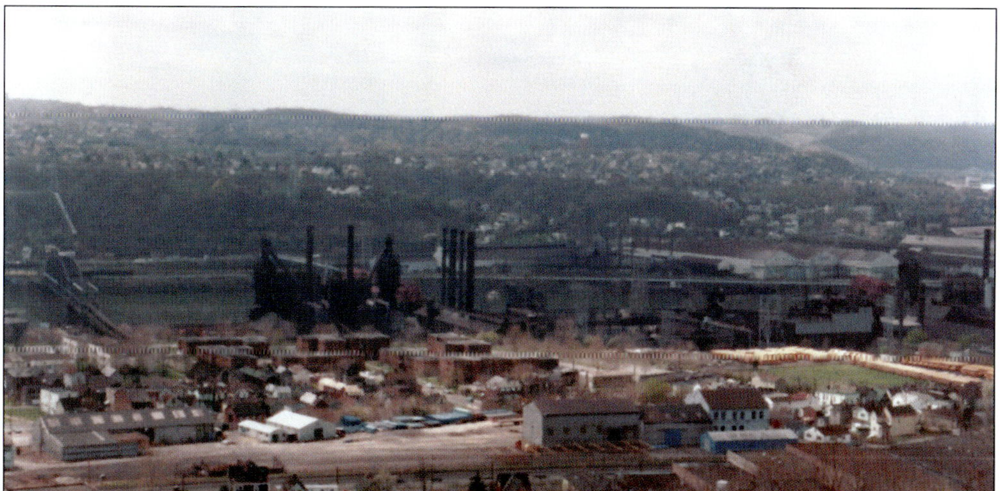

Here is an overview of the Carrie Furnace site, part of the US Steel Homestead Works, on the north side of the Monongahela River. The town of Rankin is in the foreground, and from left to right are the hot metal bridge, Blast Furnaces Nos. 4 and 3, six venting stacks of the turbo generator boiler house, Powerhouse No. 3, and Blast Furnaces No. 6 with central venting stack.

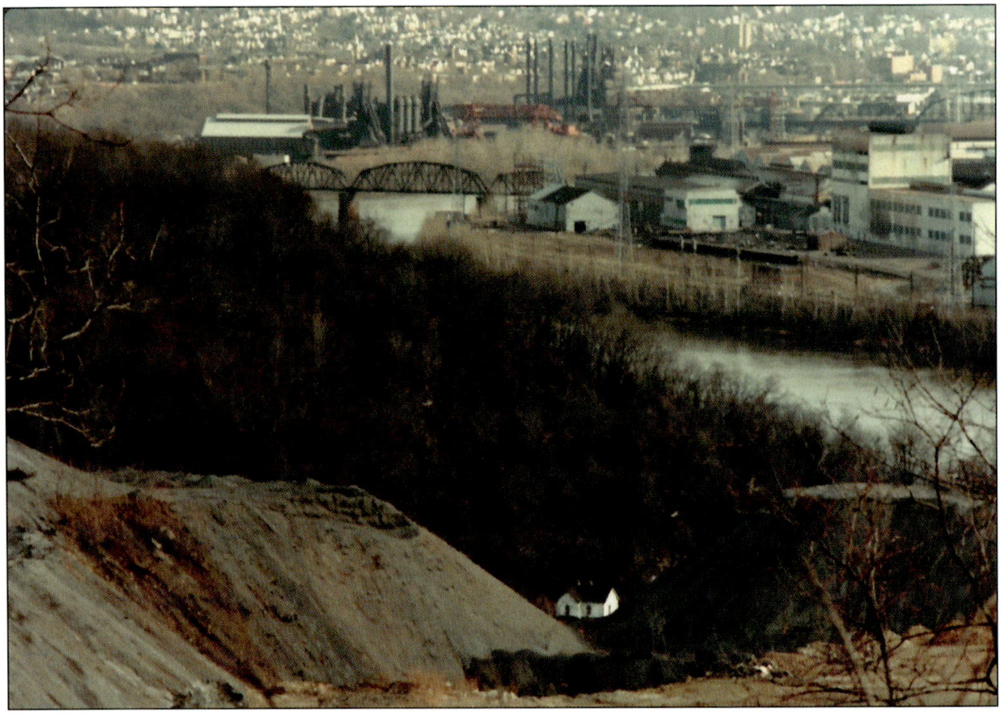

Looking upstream from the Slags are, from left to right, Blast Furnaces Nos. 7 and 6 with the Homestead plant across the river. The little white house is one of a dozen or so homes that make up the Duck Hollow community. Today, this slag dump has been covered over with topsoil, and new housing has been built on it called Summerset at Frick Park.

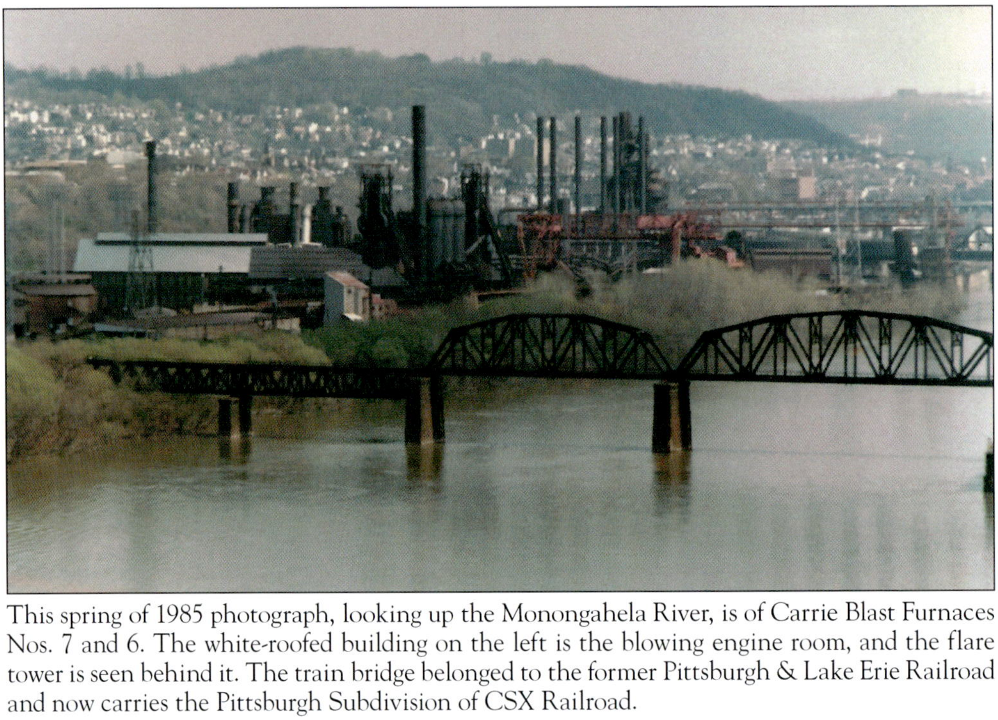

This spring of 1985 photograph, looking up the Monongahela River, is of Carrie Blast Furnaces Nos. 7 and 6. The white-roofed building on the left is the blowing engine room, and the flare tower is seen behind it. The train bridge belonged to the former Pittsburgh & Lake Erie Railroad and now carries the Pittsburgh Subdivision of CSX Railroad.

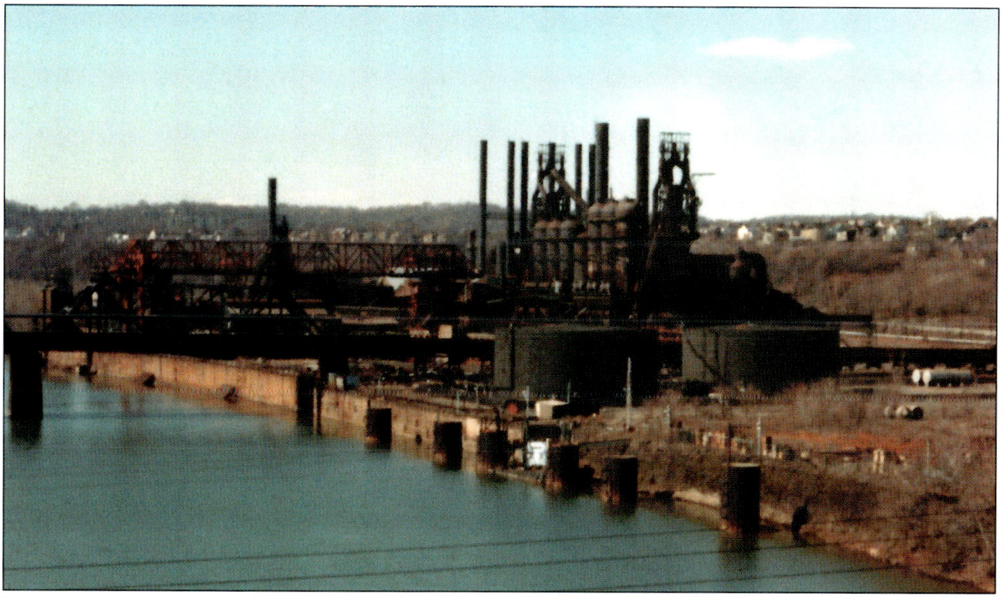

Looking downriver from the Rankin Bridge are the Carrie Blast Furnaces Nos. 4 and 3, built in 1941. The hot metal bridge that crosses the Monongahela River opened in 1900. The bridge took molten iron from Carrie Furnace over to the open-hearth furnaces at Homestead to be refined into steel. The sides of the bridge were shielded to prevent splashing metal from hitting the water and exploding.

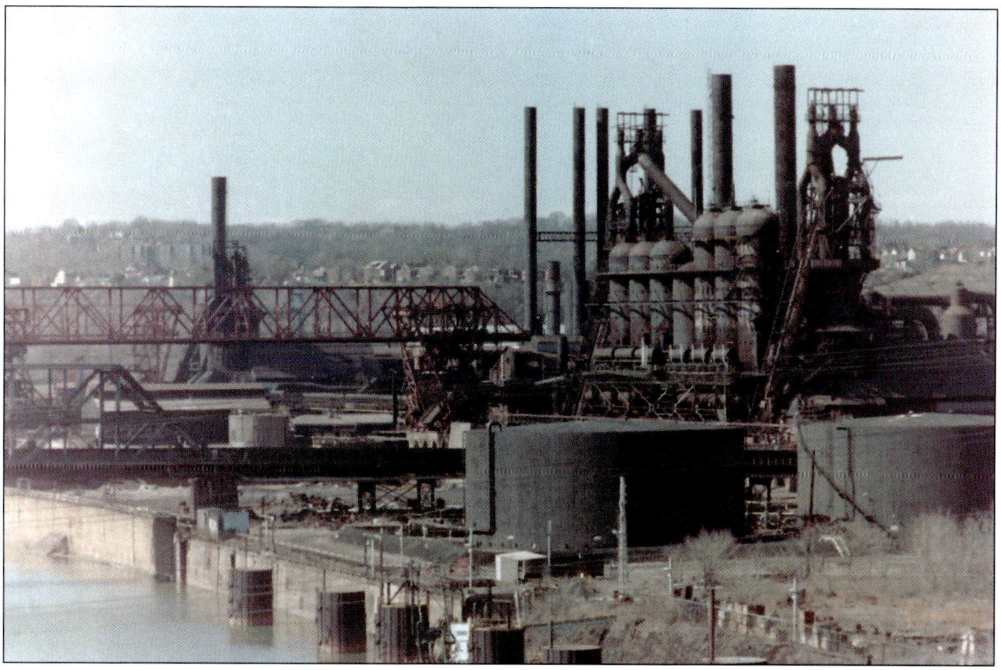

This July 1985 photograph is a closer look at Blast Furnaces Nos. 4 and 3. Two water tanks are seen in the foreground, along with the hot metal bridge and 12-ton ore bridge behind. Six draft stacks for the turbo generator boiler house and Blast Furnaces Nos. 6 and 7 are in the background.

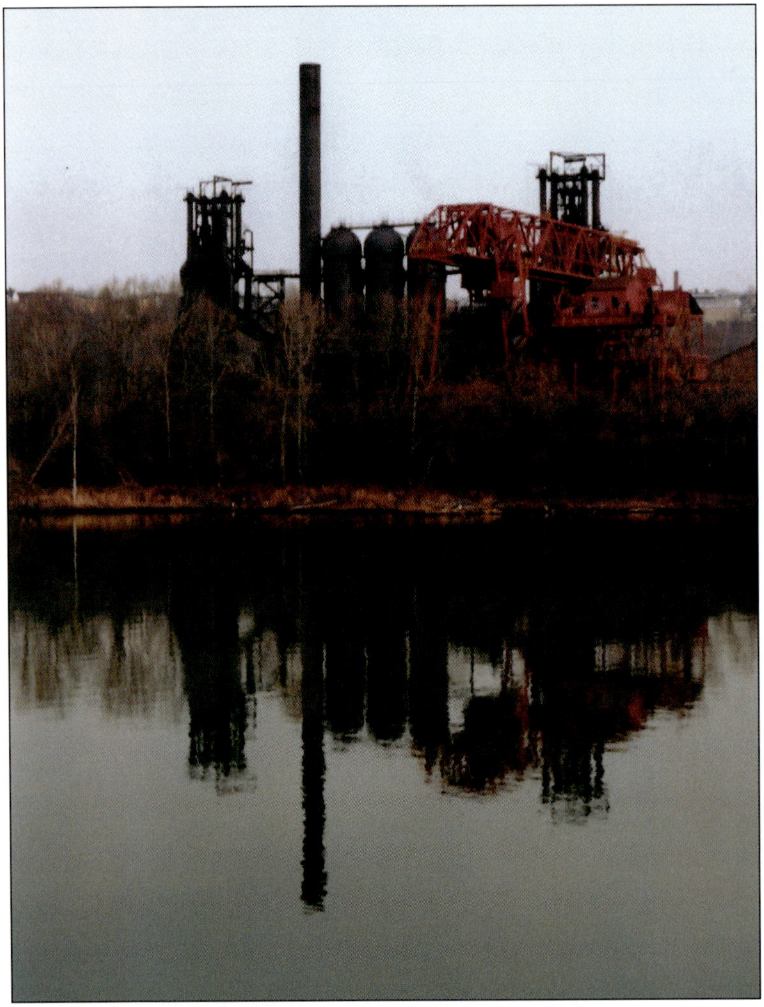

This view of Carrie Furnace shows, from left to right, Blast Furnaces No. 7 and No. 6 (left to right) reflected in a still Monongahela River. The 7 1/2-ton capacity ore bridge is on the right. Trees have grown up in the last 25 years that were not there when the furnace was active. This photograph was taken around November 1992.

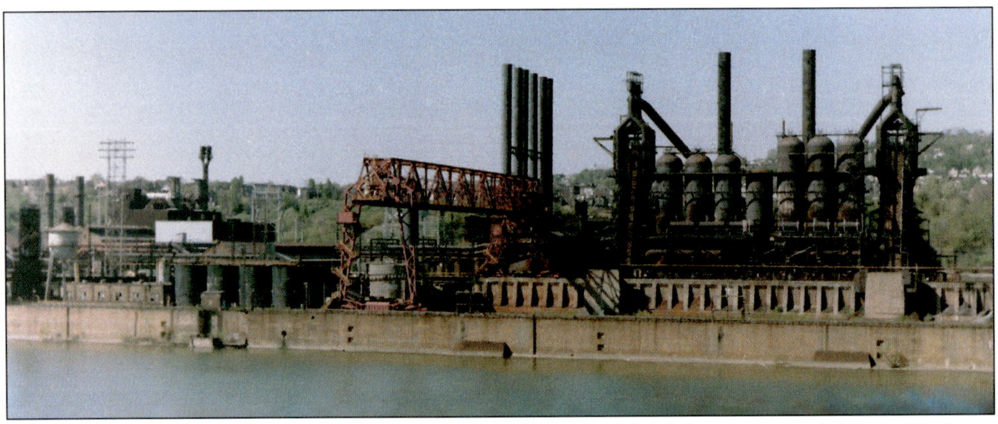

Looking north across the river are Blast Furnaces Nos. 3 and 4. They were built by the Defense Plant Corp. in 1941. The draft stacks are for the dozen 1,112-horsepower turbo power generators in the boiler house. Coal was burned to heat water from the river to make steam to run theses behemoth generators. The tanks to the left were part of the water purifying plant.

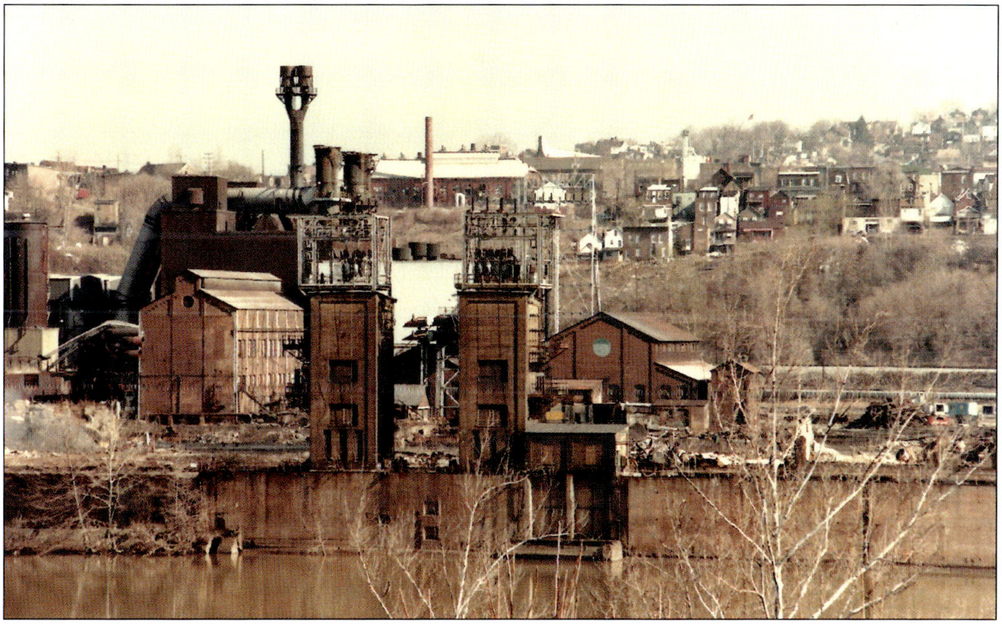

This view from across the Monongahela River shows the middle section of the Carrie Furnace plant. One can see two transformer stations (the tall structures at center) that used to send electric power over to Homestead. To the right is the intake pier with a tunnel where water was pumped to the boiler house. A double-headed flare tower is seen in the background.

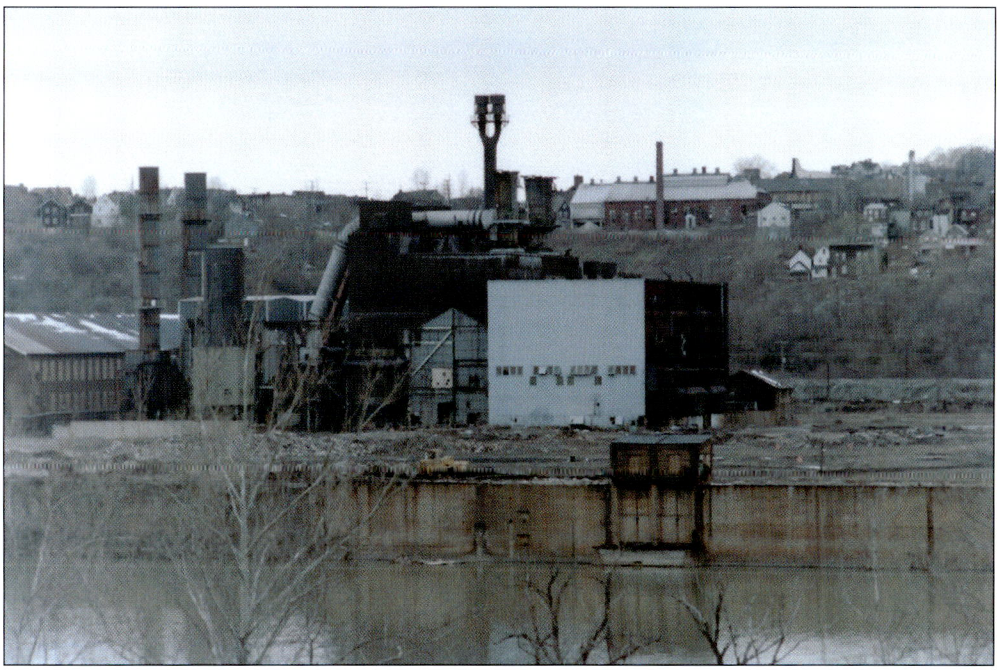

This is the same general view as the previous photograph of the middle section of the Carrie Furnace site taken a year later after some buildings have been removed. The white-sided building was Powerhouse No. 3.

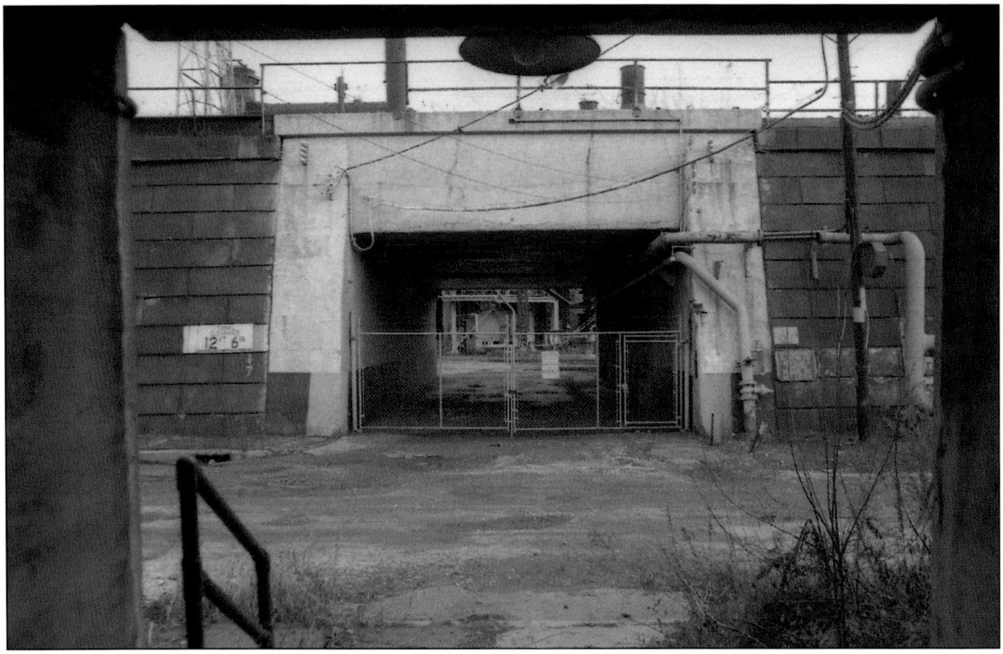

This black-and-white print, taken in December 1984, is one of the main employee entrance gates. The gate gives access to the middle part of the plant and goes under the double set of Union Railroad tracks that run on top of the wall that goes along this side of the site. The gate has been reopened and is used today by the Rivers of Steel group to bring tour groups in.

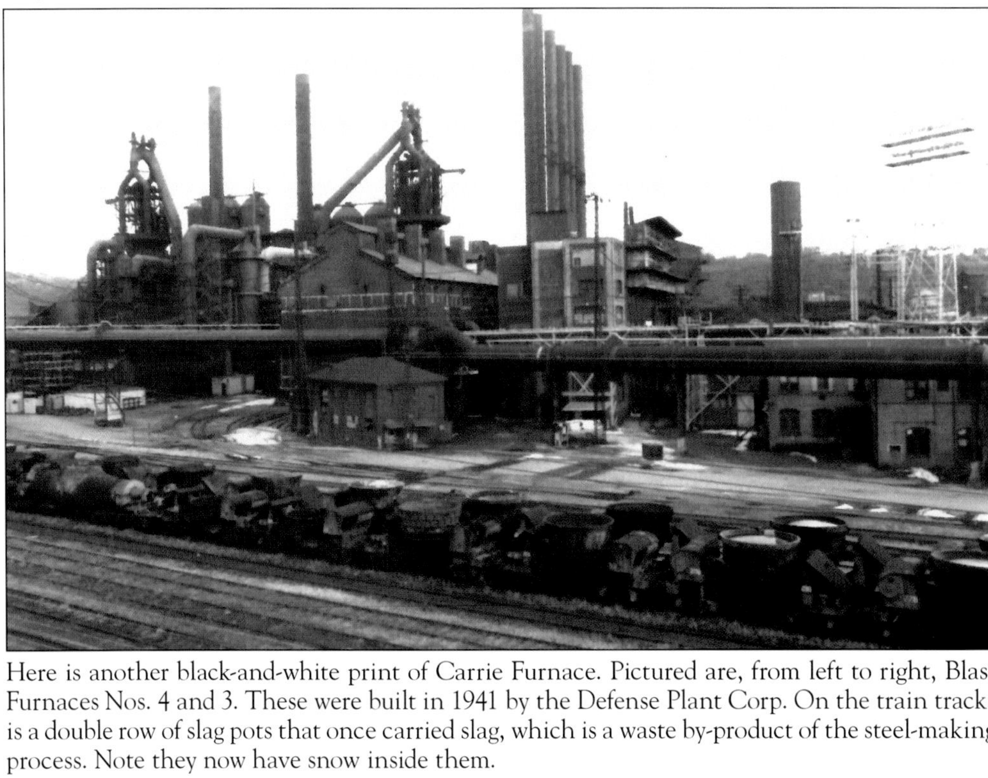

Here is another black-and-white print of Carrie Furnace. Pictured are, from left to right, Blast Furnaces Nos. 4 and 3. These were built in 1941 by the Defense Plant Corp. On the train tracks is a double row of slag pots that once carried slag, which is a waste by-product of the steel-making process. Note they now have snow inside them.

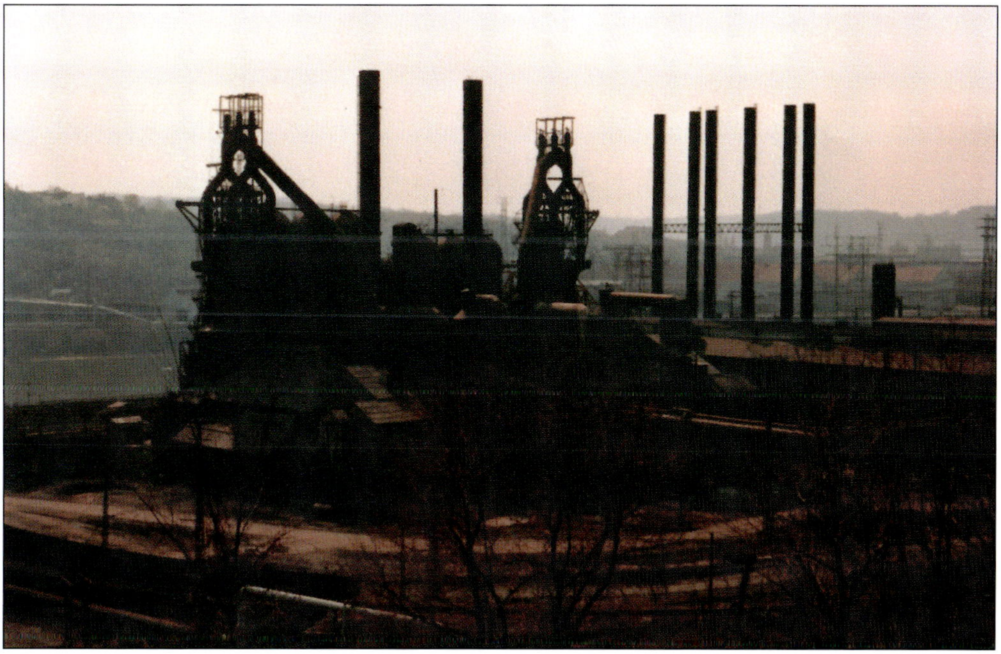

From left to right, Carrie Blast Furnaces Nos. 4 and 3 are seen in the late autumn of 1985 in this southwest-looking view from up on the hill in Rankin, Pennsylvania. Built in 1941, the furnaces had the capacity to make 1,300 tons of molten iron per day. These furnaces made the iron that made the steel that won the war to make the world safe for democracy.

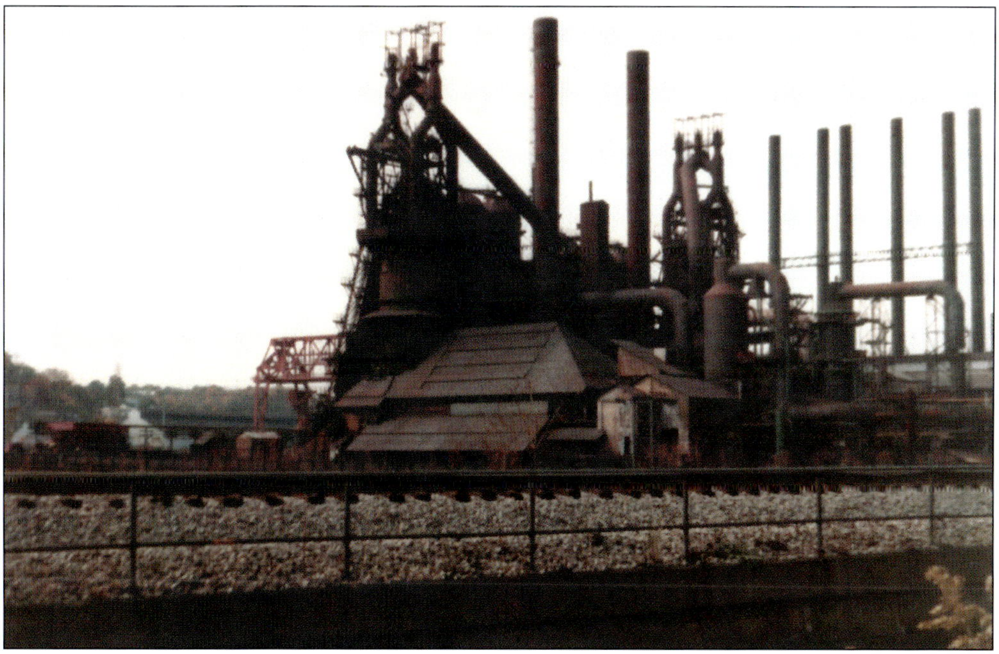

Another view of Blast Furnaces Nos. 4 and 3 shows a clear view of Blast Furnace No. 3's cast house. These "high-tops" were built in 1941 and demolished around 1990. The photograph was taken from the B&O main line tracks looking over the P&LE Railroad tracks that run along the top of the wall.

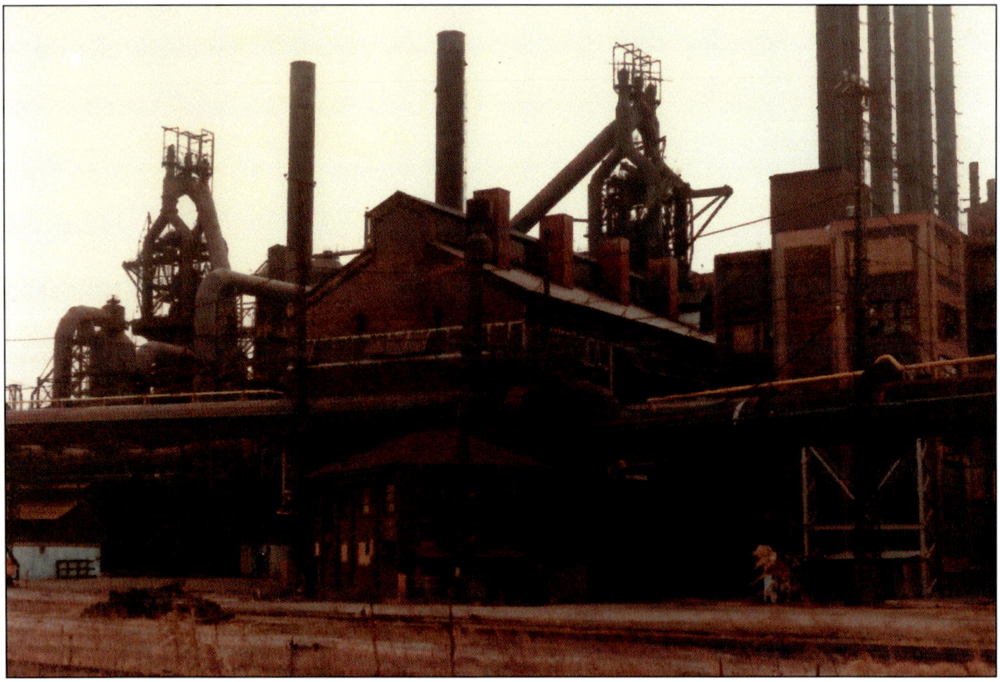

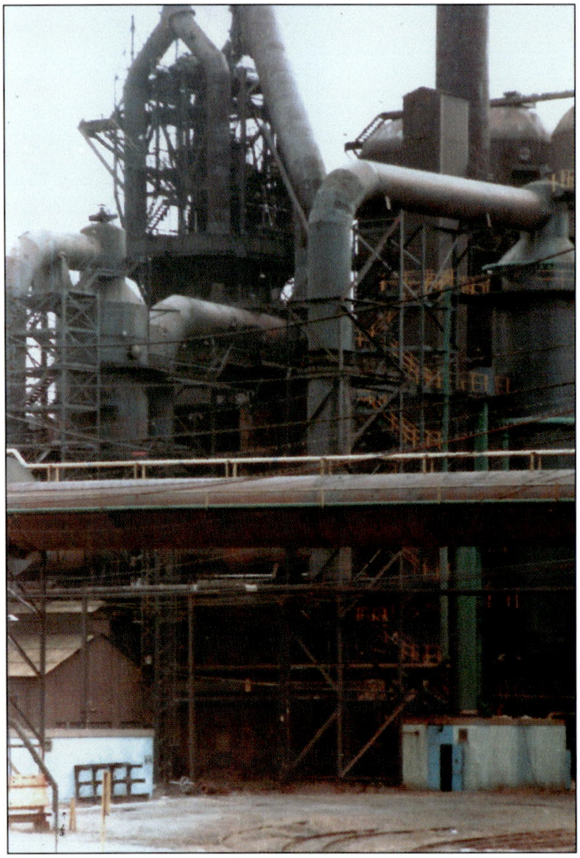

Walking up to Blast Furnaces Nos. 4 and 3 (from left and right), one can get an idea of how big they were by how they dwarfed the two-story building in the foreground. When this photograph was taken in 1985, the furnaces were still pretty much as they had been left on the last day of work.

Pictured is Blast Furnace No. 4. At the time of this photograph, a person caught on the premises of the plant would face trespassing charges and a fine.

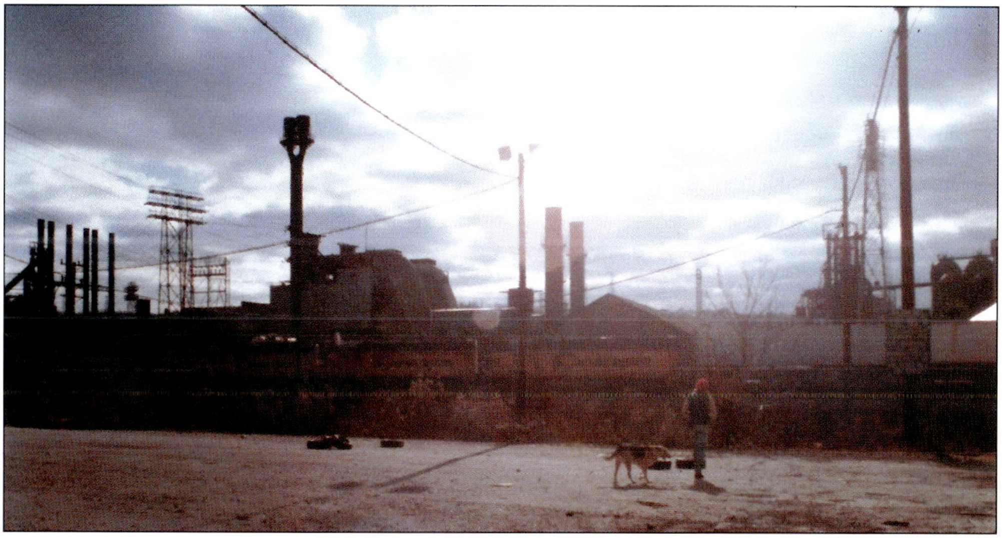
This photograph, taken on Thanksgiving Day 1984, is of the empty employee parking lot, which is today used by kids riding dirt bikes.

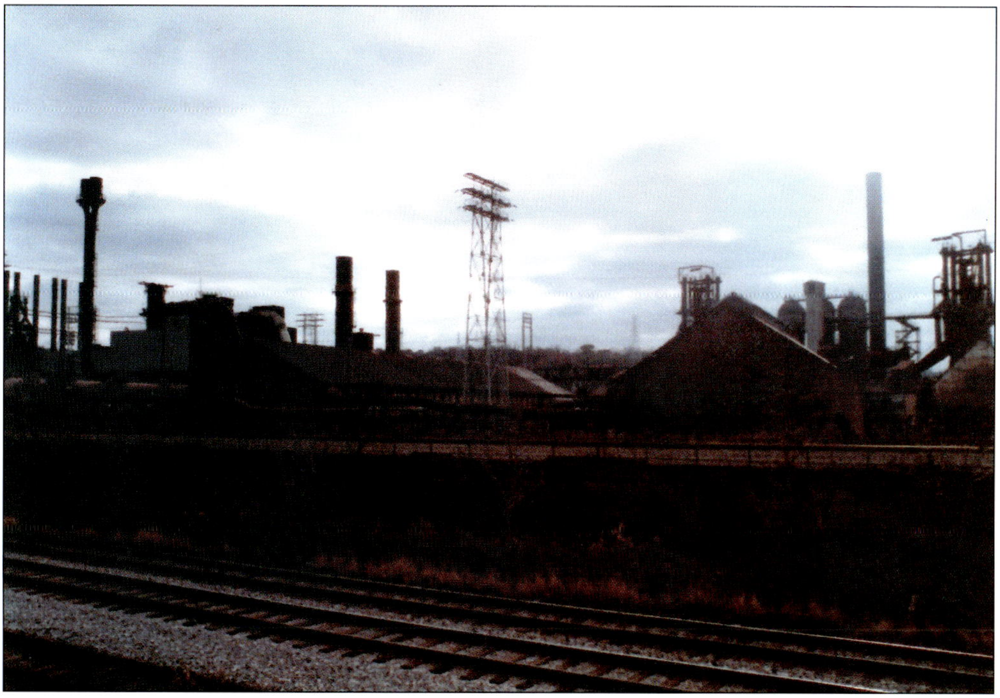
Here is another photograph taken on Thanksgiving Day 1984; it is a more western view of Carrie Furnace. The flare tower that burned off excess natural gas is on the left; an electrical tower is seen in the center of the photograph. Blast Furnaces Nos. 6 and 7 are to the right. The elevator shaft, Cowper hot-blast stoves, and draft stack are seen between the furnaces.

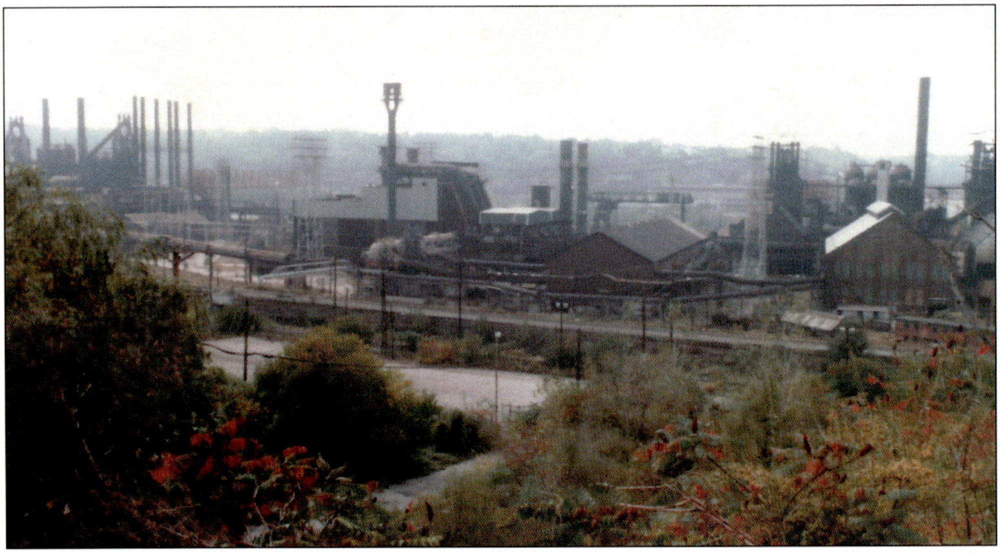

This c. 1985 view is of the north side of Carrie Furnace before the demolition had begun. This photograph was taken from the backyard of a home just up the hill in Swissvale. The employee parking lot is seen in the foreground. From left to right are Blast Furnaces Nos. 4 and 3, then the flare tower, and Blast Furnaces Nos. 6 and 7, with the lowing engine room between them.

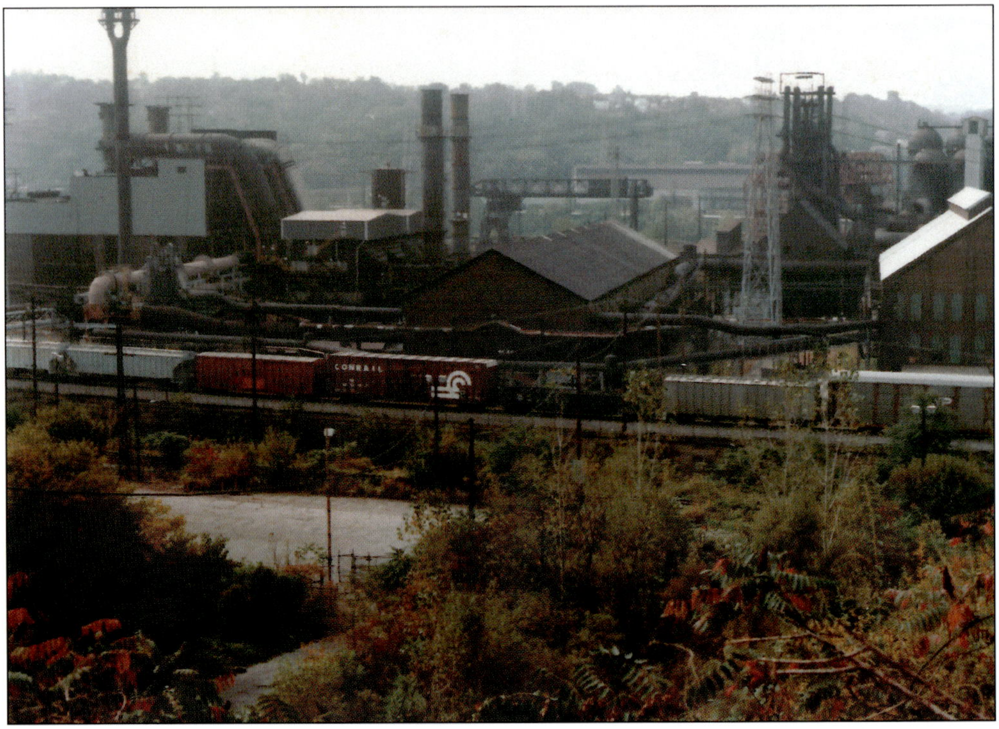

This photograph shows the middle section of the plant still intact with all the electric power and gas pipes in place. A train is going by heading west. The Conrail marking is seen on one of the cars.

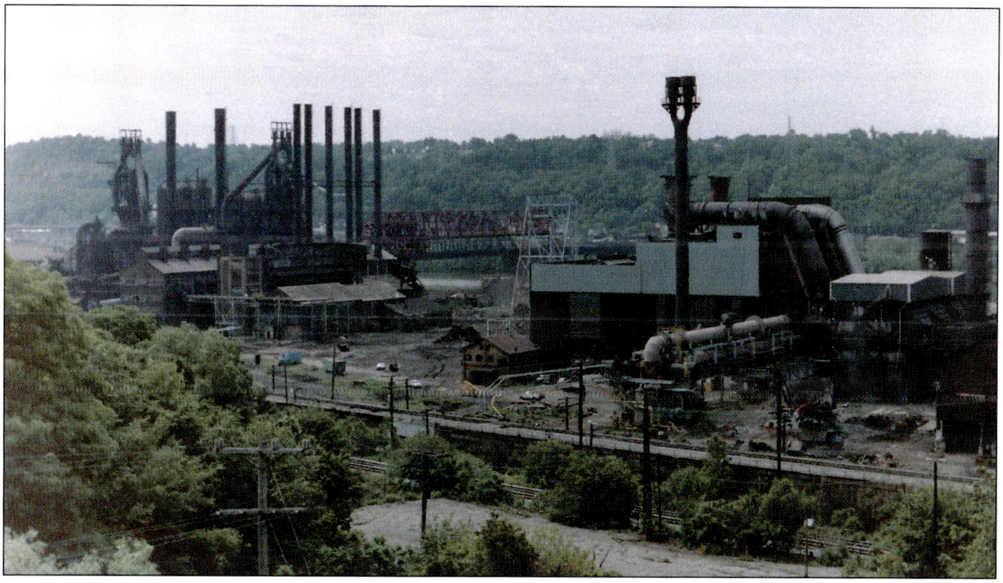

In this photograph, looking southwest, one sees, from left to right, Carrie Blast Furnaces Nos. 3 and 4. The yellow-brick building (center) was the plant's hospital. The double-headed flare tower, just to the right, burns off excess natural gas. A little farther to the right a backhoe is seen loading a truck with debris from the buildings.

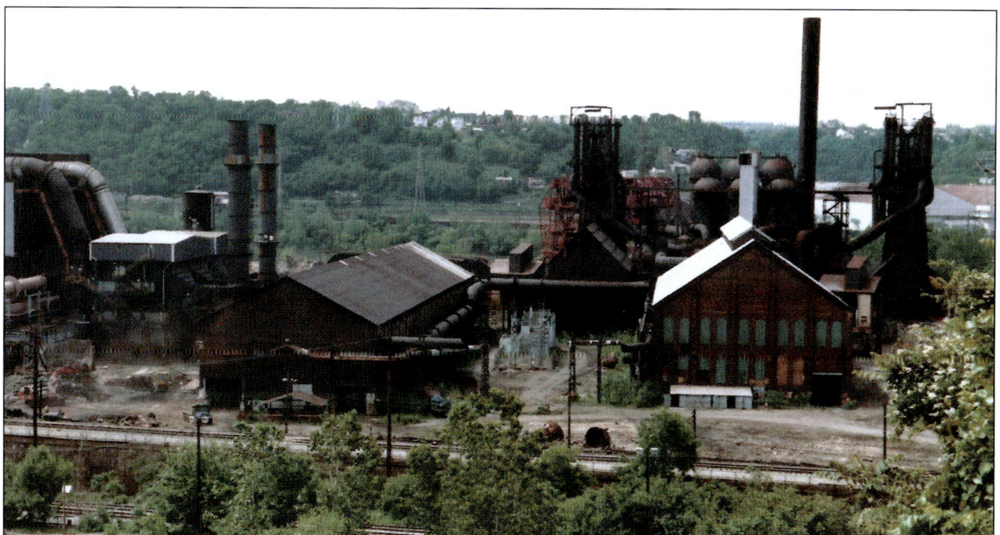

On the left, the backhoe is still at work. To the right are Blast Furnaces Nos. 6 and 7. The electrical tower and gas pipes have already been demolished. The white-roofed building between the furnaces is the blowing engine room.

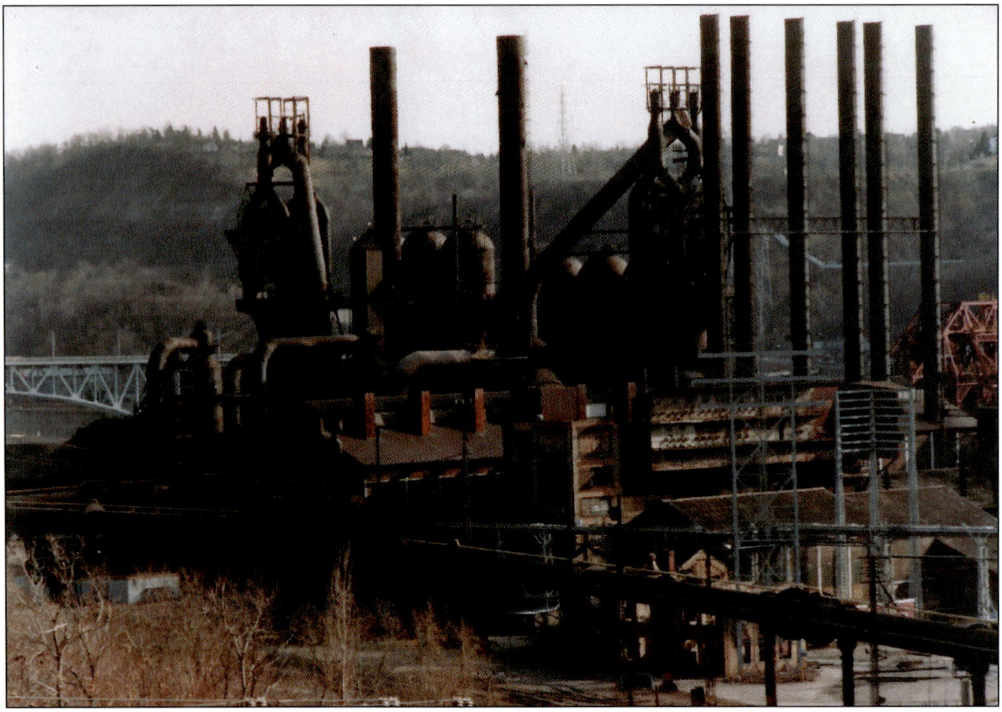

Part of the gearing up for war, the Blast Furnaces Nos. 4 and 3 were built in 1941 by the Defense Plant Corp. In this late 1985 autumn view, they were still untouched.

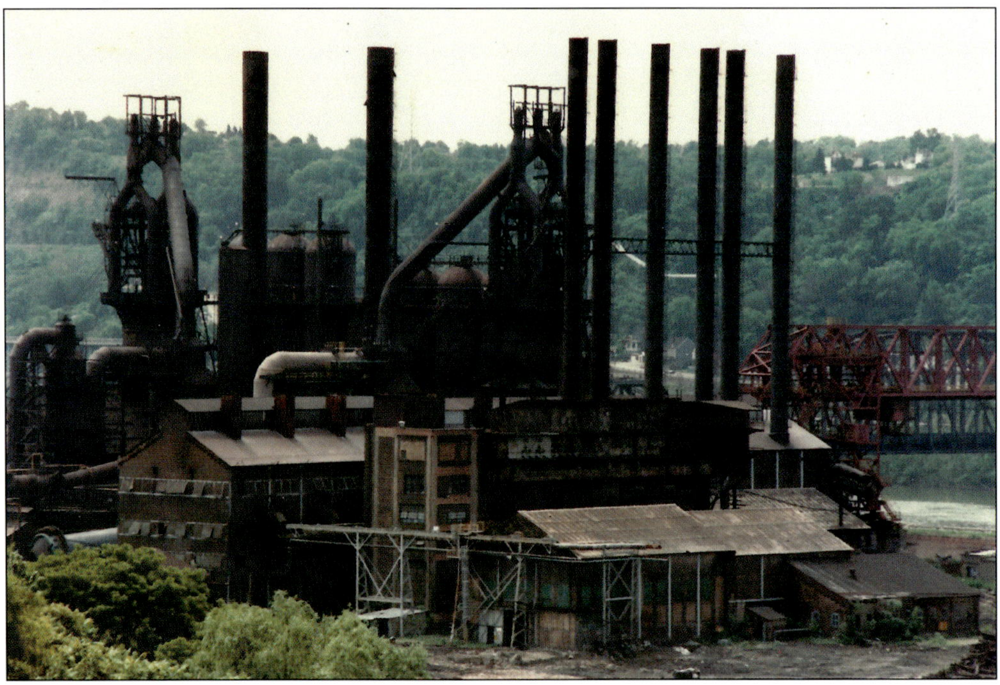

Here is another photograph of the blast furnaces taken the following spring from the same vantage point. As one can see, some work has been done demolishing parts of the site. The eight-foot, six-inch-diameter clean gas pipe is gone, along with several other buildings.

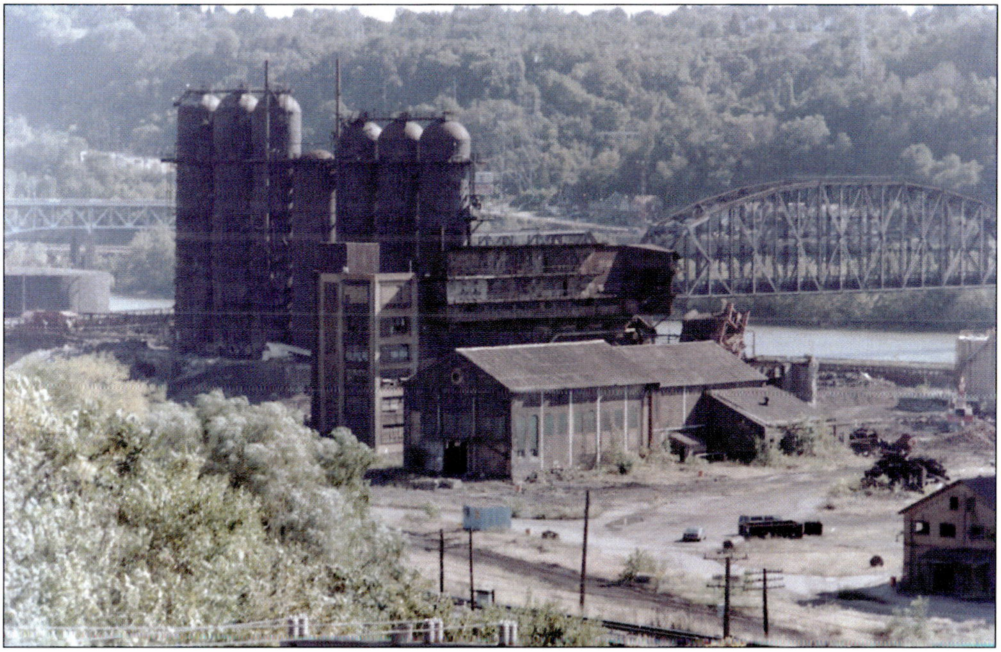

By this time, the two blast furnaces, the ore bridge, and much more have all disappeared. The Cowper hot-blast stoves, the turbo generation station, and the pump house with the carpenters' shop along its side are still standing. One can see the hot metal bridge a lot better in this photograph.

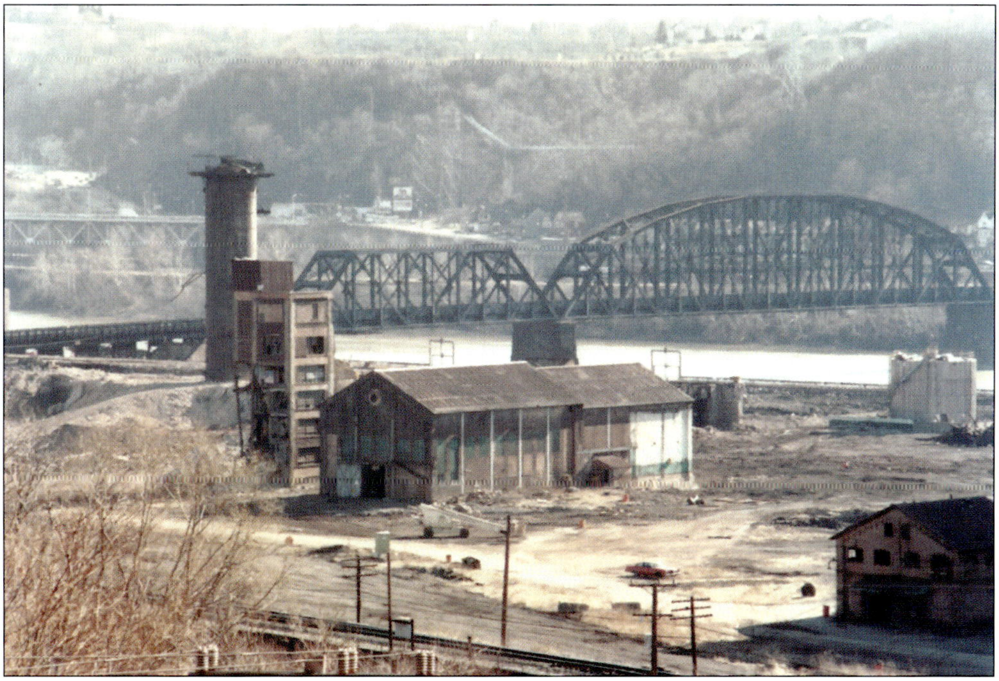

By February 1991, only the center part of the Cowper hot-blast stoves still stood with the southern end of the turbo generation station. The pump house is seen, but the carpenters' shop is gone. In the lower-right corner, the yellow-brick hospital building still remains.

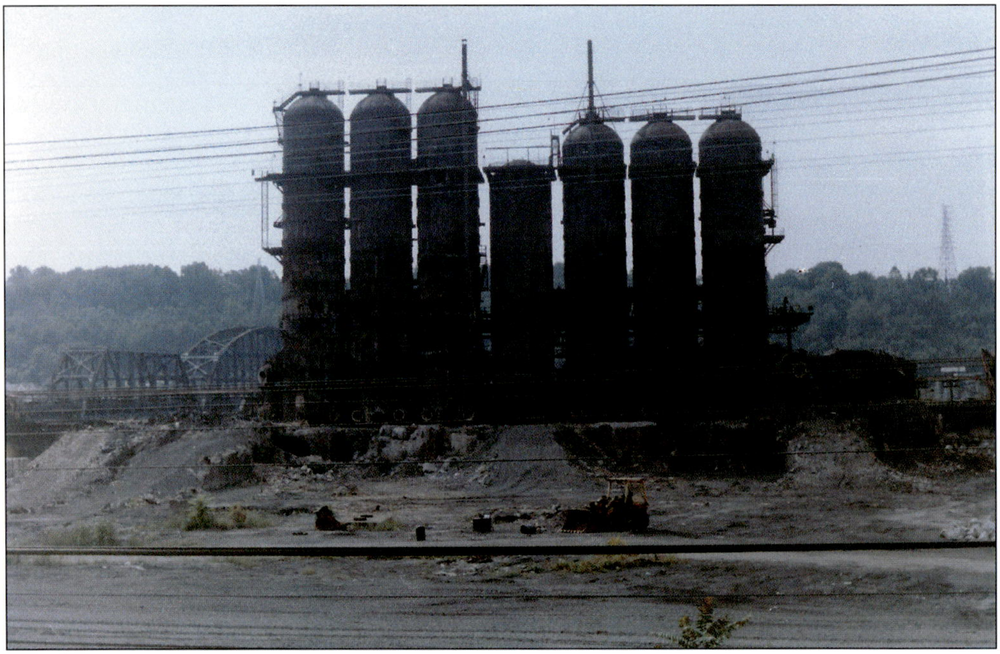

This series shows the demolishing of the Cowper hot-blast stoves. They were integral to Blast Furnaces Nos. 4 and 3 of the Carrie Furnace site. They are viewed from the north, looking south. They stand alone without the furnaces. A yellow bulldozer stands before them, ready to finish the job around November 1990.

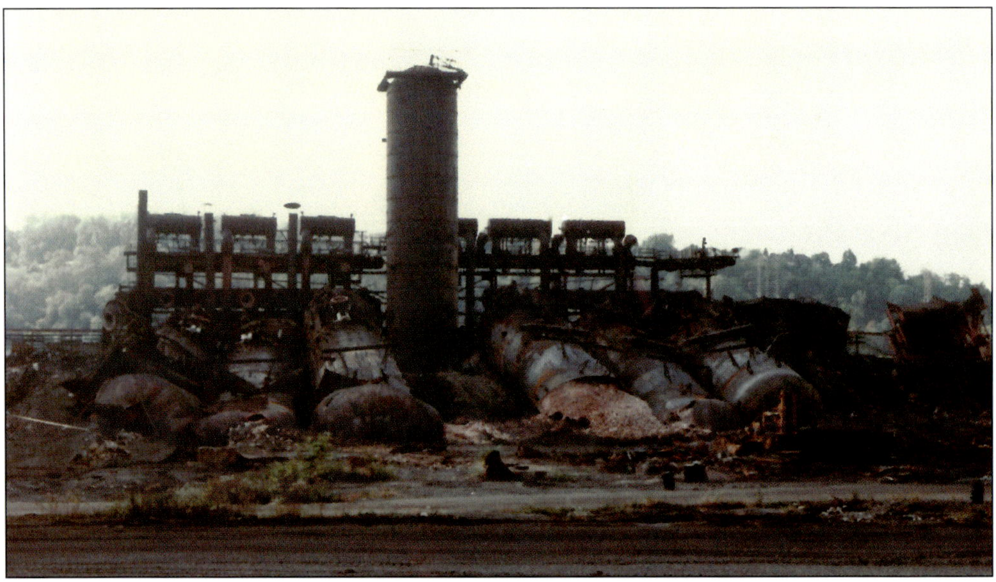

Around December 1990, the hot-blast stoves have been toppled, and their refractory brick insides have spilled out onto the ground. Only a center part remains standing.

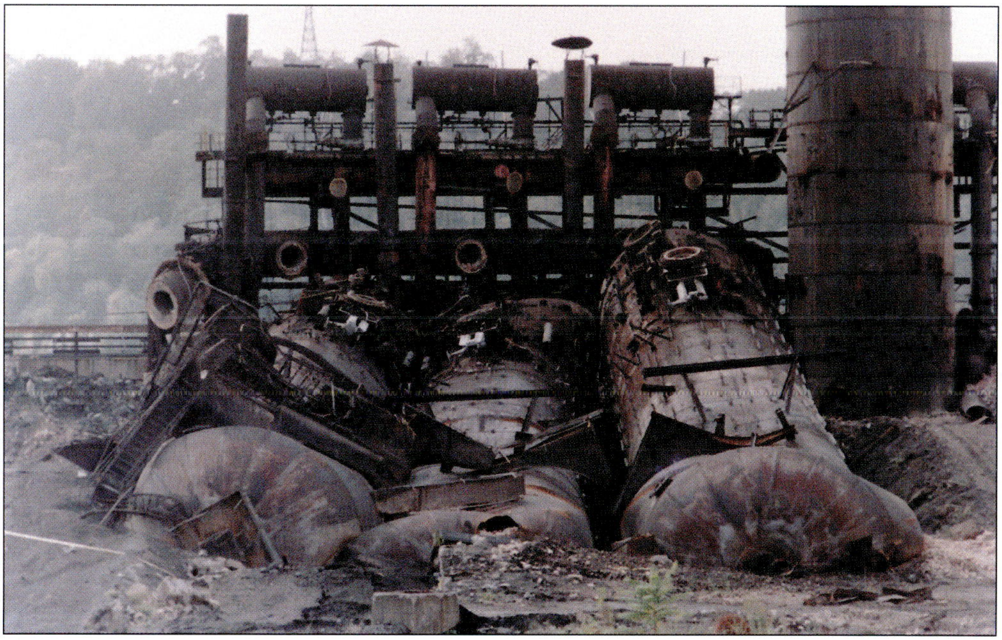

Shown the day after an explosion knocked them over in early 1990, the three Cowper hot-blast stoves on the left side operated for the Blast Furnace No. 4. They were later further demolished and trucked away as scrap metal.

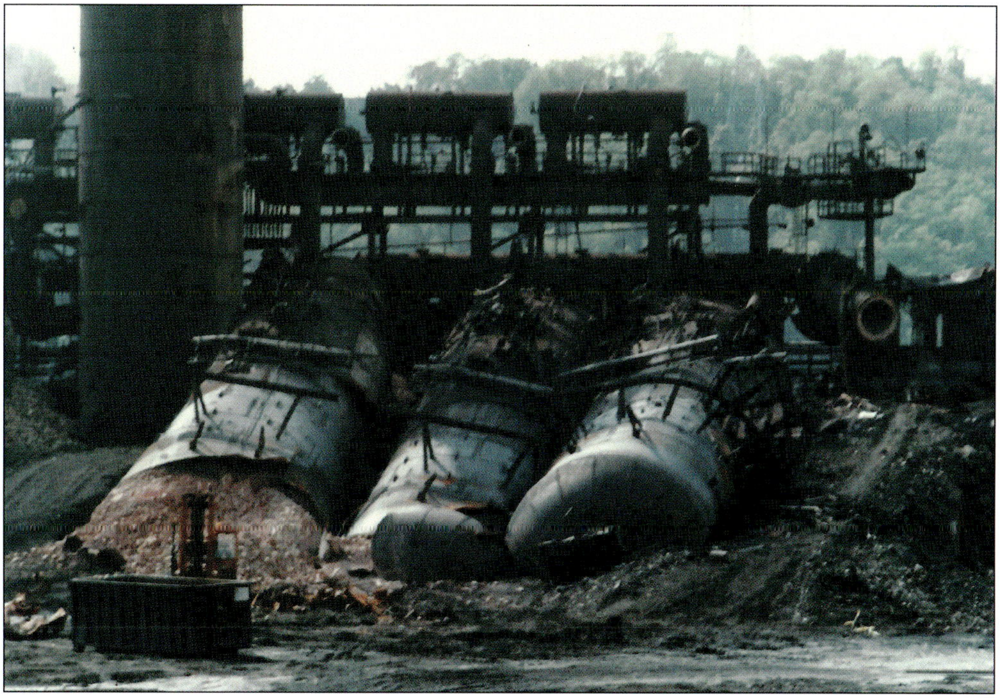

The three Cowper hot-blast stoves on the right side operated for the Blast Furnace No. 3. All were built in 1941 to make iron for the open-hearth furnaces at Homestead, Pennsylvania, just across the river. They were to help make steel for the defense of democracy against the Axis powers in World War II.

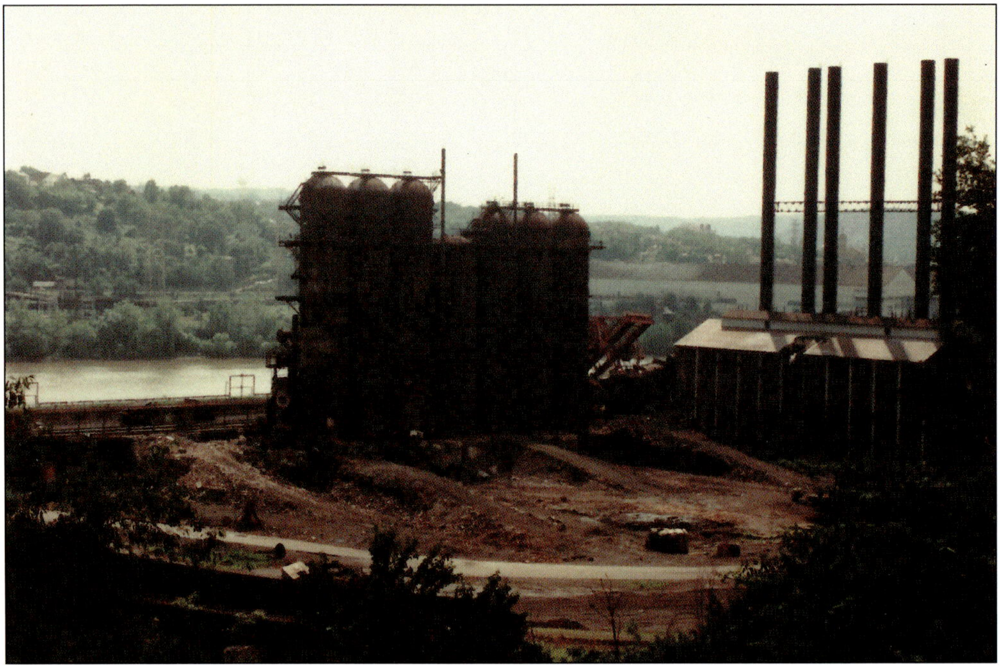

Photographed are, from left to right, Carrie Blast Furnaces Nos. 4 and 3 right after they had been demolished; the Cowper hot-blast stoves still stand along with the draft stacks of the turbo power generator house to the right.

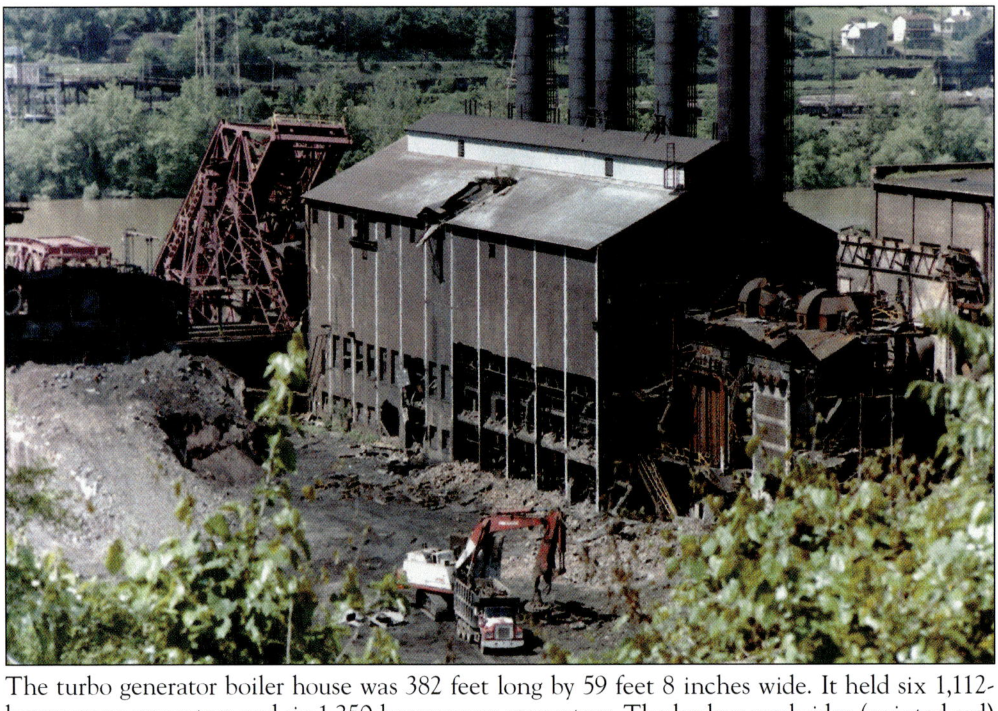

The turbo generator boiler house was 382 feet long by 59 feet 8 inches wide. It held six 1,112-horsepower generators and six 1,250-horsepower generators. The broken ore bridge (painted red) is seen lying behind the building near the river. The backhoe uses a magnet to pick up scrap metal to fill the truck. This photograph was taken around 1990.

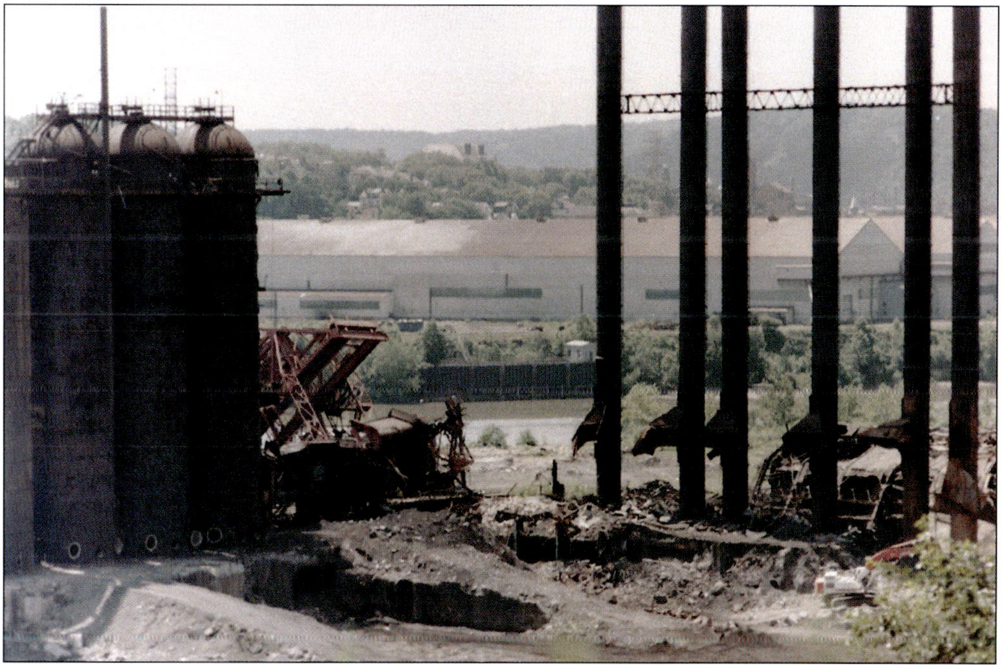

Here, one can see the draft stacks still standing after the turbo generator boiler house has been demolished. The Blast Furnace No. 3 is gone, but the Cowper hot-blast stoves still stand. The red-painted ore bridge lies in ruins, and one can see across the Monongahela River.

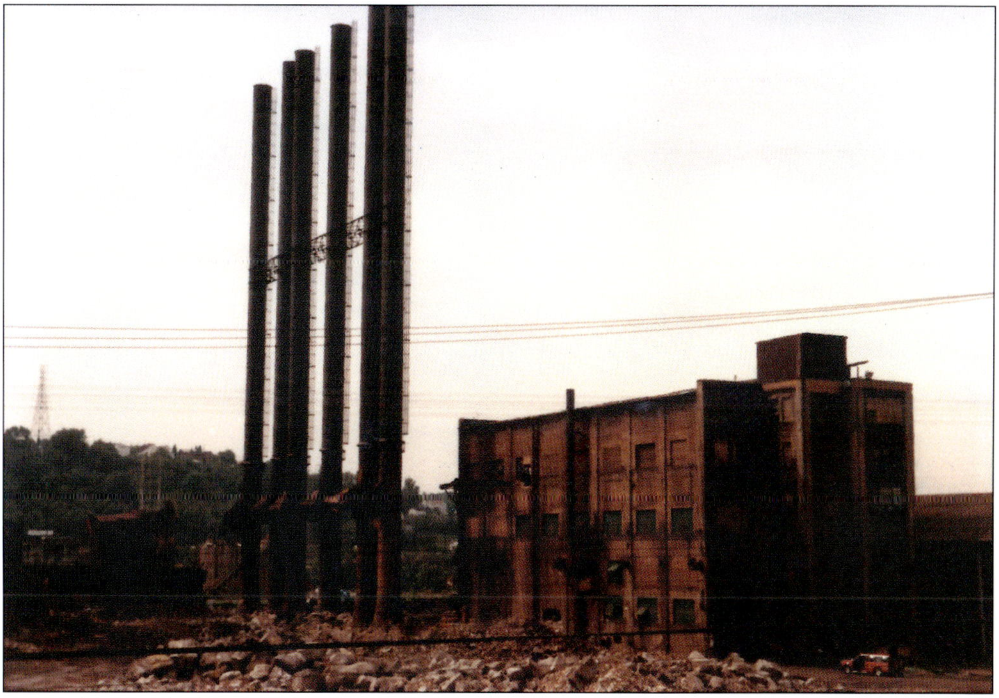

Here, the six draft stacks for the turbo generator house still stand after the building has been demolished. Next to them is the turbo generator station, which was 252 feet long by 66 feet wide.

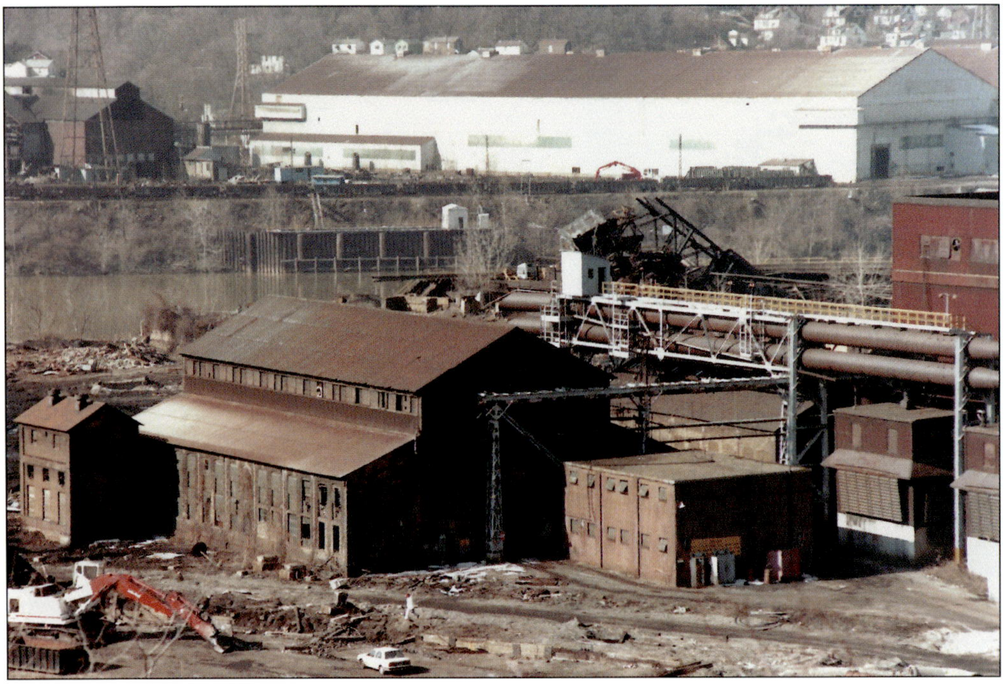

In March 1989, the Carrie Furnace machine shop is seen here with a backhoe in the lower-left corner of the photograph. A white automobile is parked a little to the right, and a man running back to his machine is there just above the car.

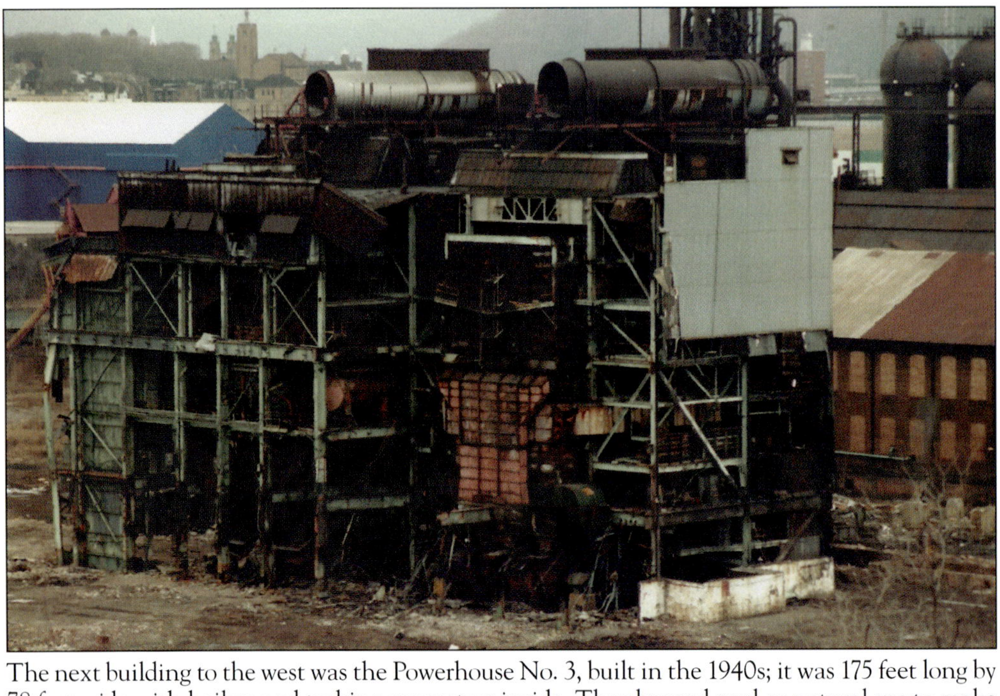

The next building to the west was the Powerhouse No. 3, built in the 1940s; it was 175 feet long by 78 feet wide with boilers and turbine generators inside. They burned coal or natural gas to make steam to run the turbos to make electricity. In this photograph, one can see that the buildings whole east end has been ripped off, showing the inside of the powerhouse.

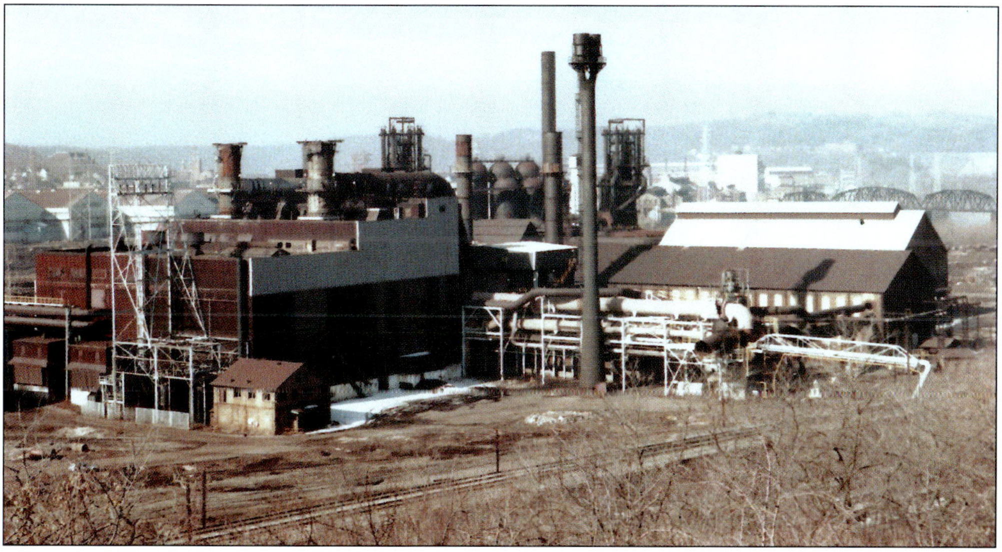

This photograph of the middle section of the plant was taken from up on a hillside, looking southwest. One can see the yellow-brick hospital building in the lower left with Powerhouse No. 3 behind it and the flare tower over on the right.

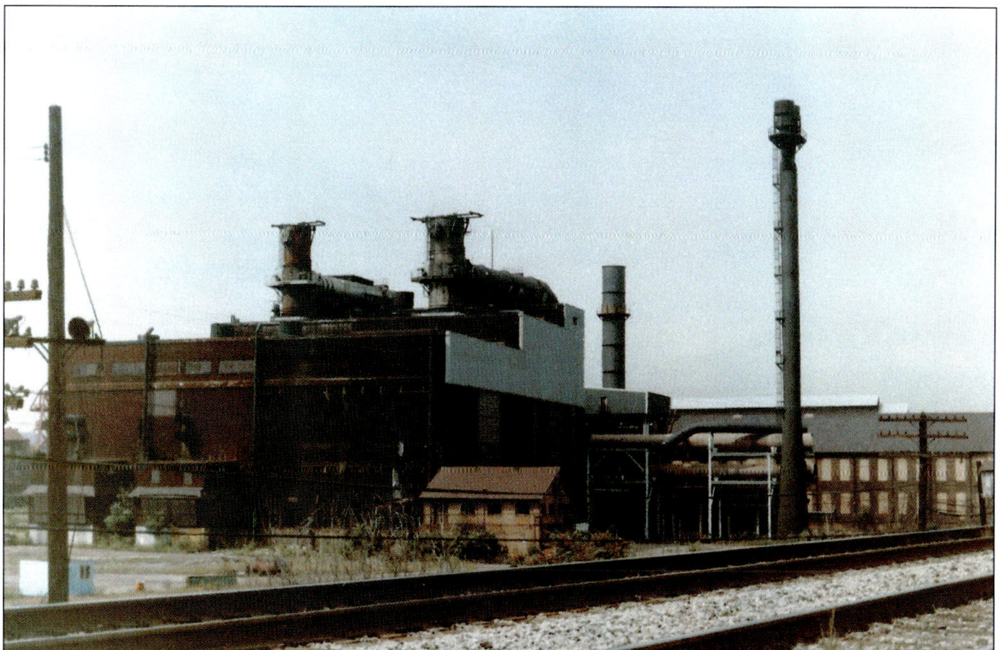

Here is a closer view of the same area, but this time, the photograph was taken from the Union Railroad tracks. Again, one can see the yellow-brick hospital building, Powerhouse No. 3 behind it, and the flare tower to the right.

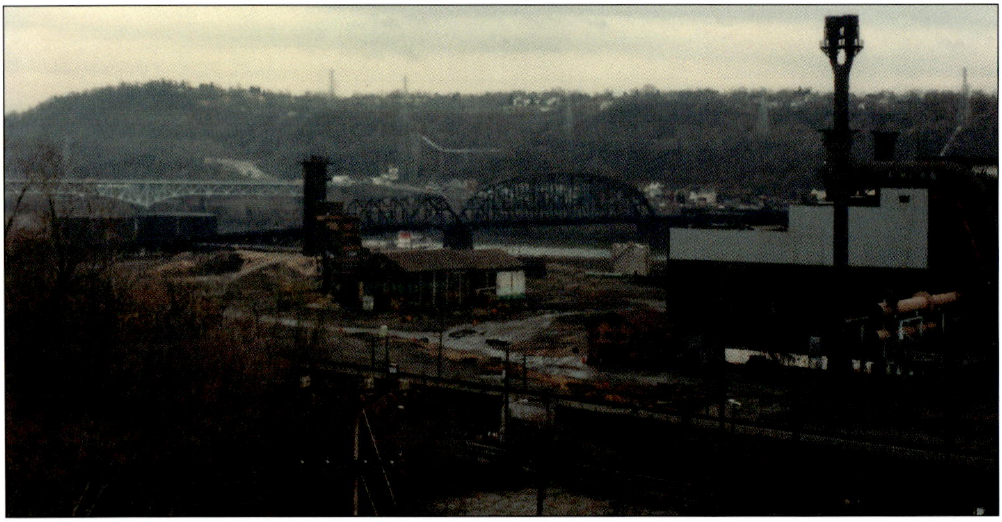

Looking southeast from a home in Swissvale, Pennsylvania, one can see the hot metal bridge over the Monongahela River in the distance. The top of the employee entrance (painted white) is visible at the bottom center of the photograph. Powerhouse No. 3 with the double-headed flare tower is in front.

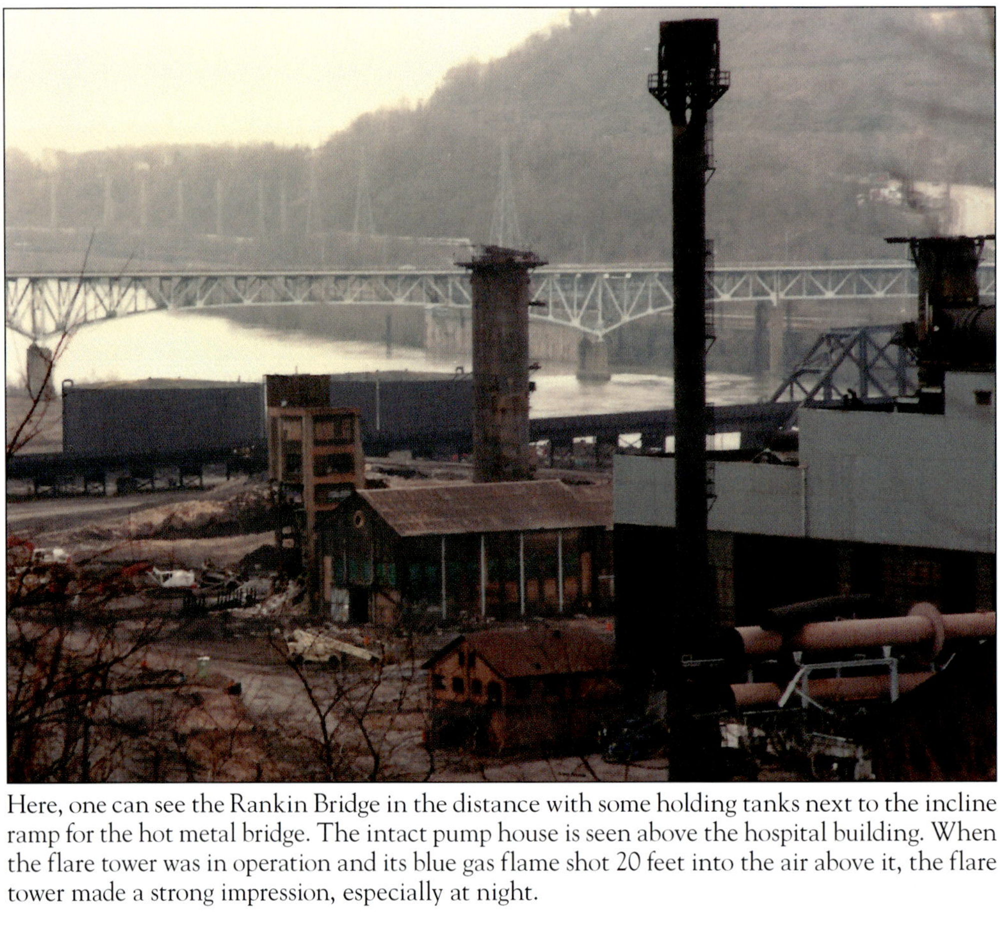

Here, one can see the Rankin Bridge in the distance with some holding tanks next to the incline ramp for the hot metal bridge. The intact pump house is seen above the hospital building. When the flare tower was in operation and its blue gas flame shot 20 feet into the air above it, the flare tower made a strong impression, especially at night.

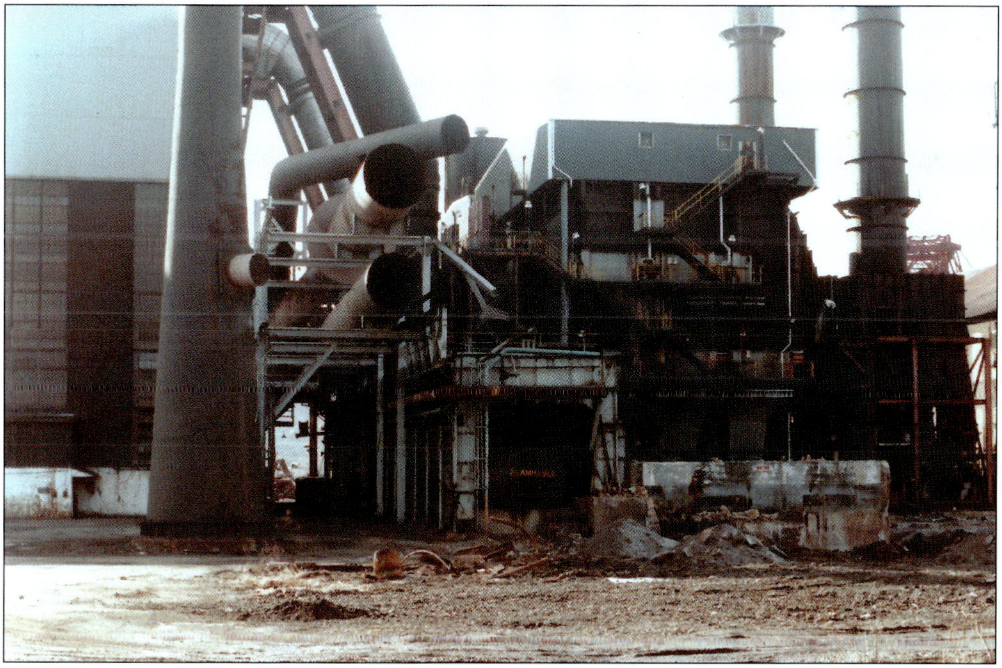

The base of the flare tower is just to the left. This was the view just inside the main employee gate during mid-demolition. The train tracks and gas pipes are already gone. All of this area has been cleared and leveled now.

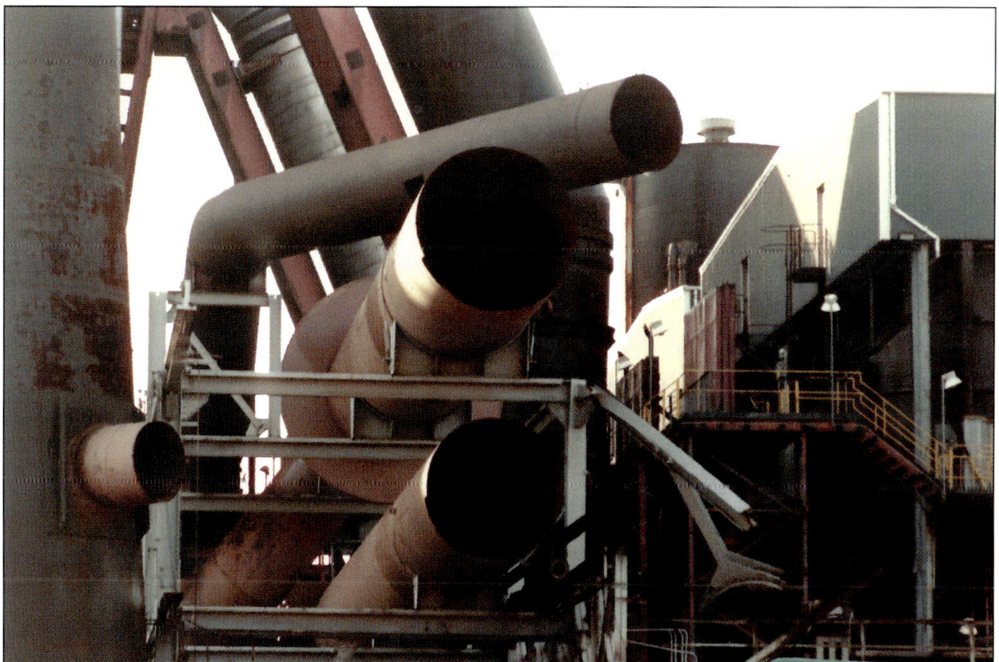

Pictured in 1991 is a close-up of the cut ends of various pipes that took the "natural gas main" or the "cold blast main" to different parts of the plant. The large pipe in the middle (natural gas main) measured eight feet, six inches in diameter.

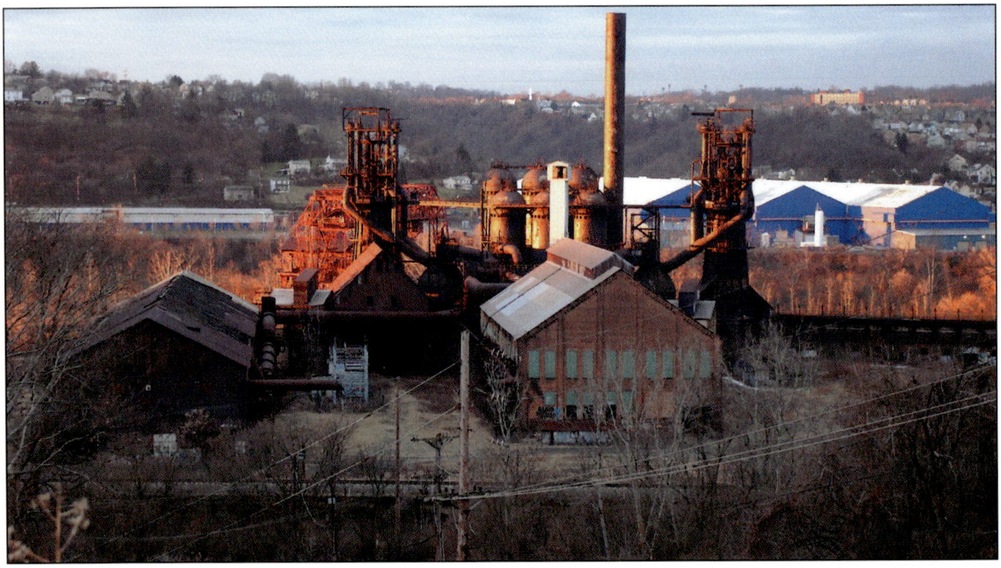

This is a more recent view of what remains of the Carrie Furnace. Taken in 2010, this is what most people think of when they hear "Carrie Furnace." The AC (alternating current) power house is on the left, and the blowing engine room is on the right. Blast Furnaces Nos. 6 and 7 are in the background. This was taken in the early morning, when the first rays of the sun coming from the east (left) touched the tops of the furnaces.

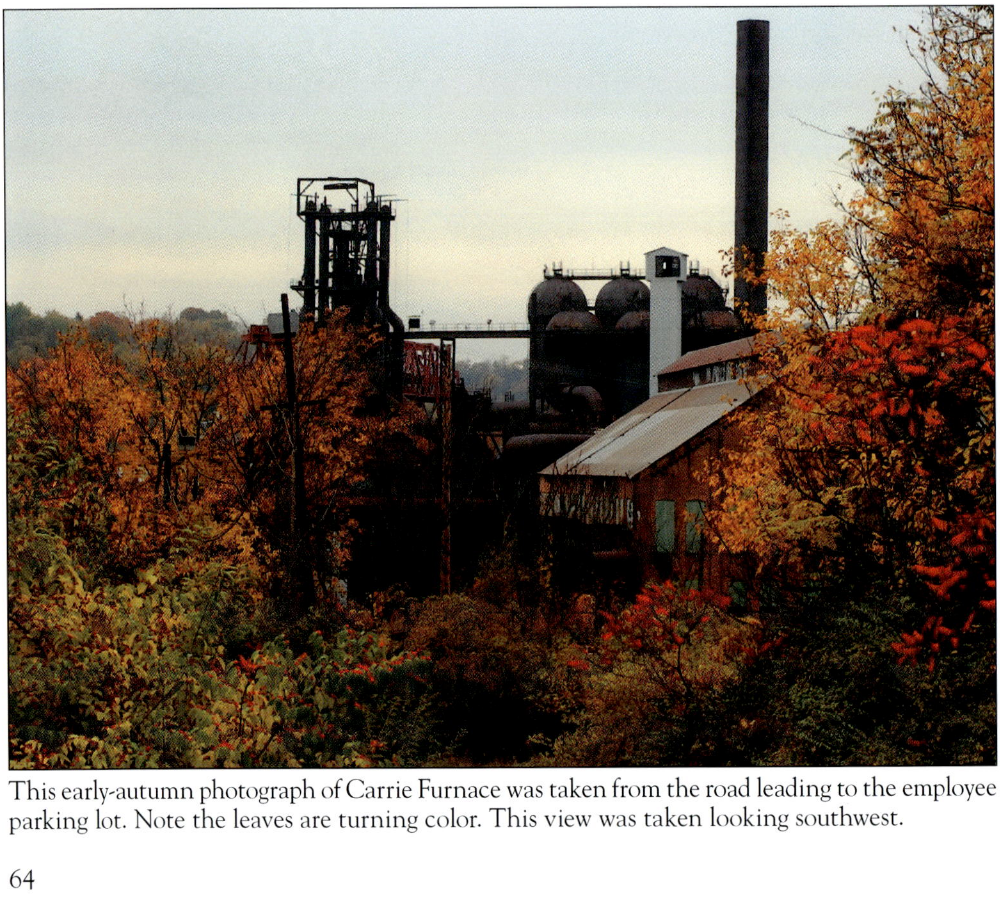

This early-autumn photograph of Carrie Furnace was taken from the road leading to the employee parking lot. Note the leaves are turning color. This view was taken looking southwest.

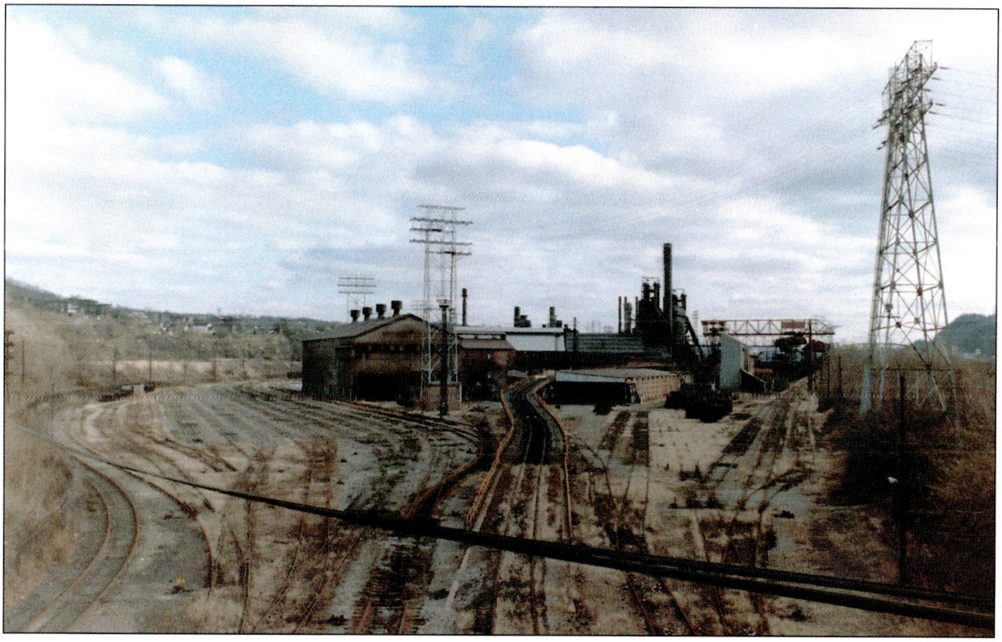

This c. 1985 overview photograph of the Carrie Furnace plant was taken from the trestle leading up to the P&LE Railroad Bridge at Munhall. The tracks and ancillary buildings were intact, and the high-voltage power lines are still standing. The center track goes to the high-line materials trestle. To the right is the ore storage yard.

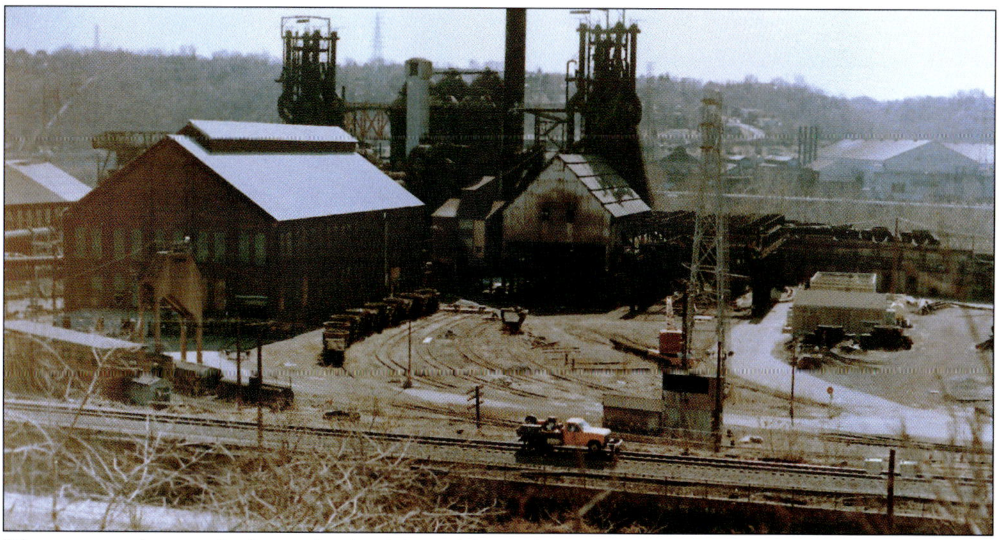

This view, taken with the author's 135-mml lens, shows a line of slag pots outside the blowing engine room and Blast Furnace No. 7 with its cast house still attached. The removal of this cast house was the last bit of demolition done by the Park Corp. of Cleveland, Ohio, at Carrie Furnace. Next, to the left of the cast house, is the hoist house for Blast Furnace No. 7.

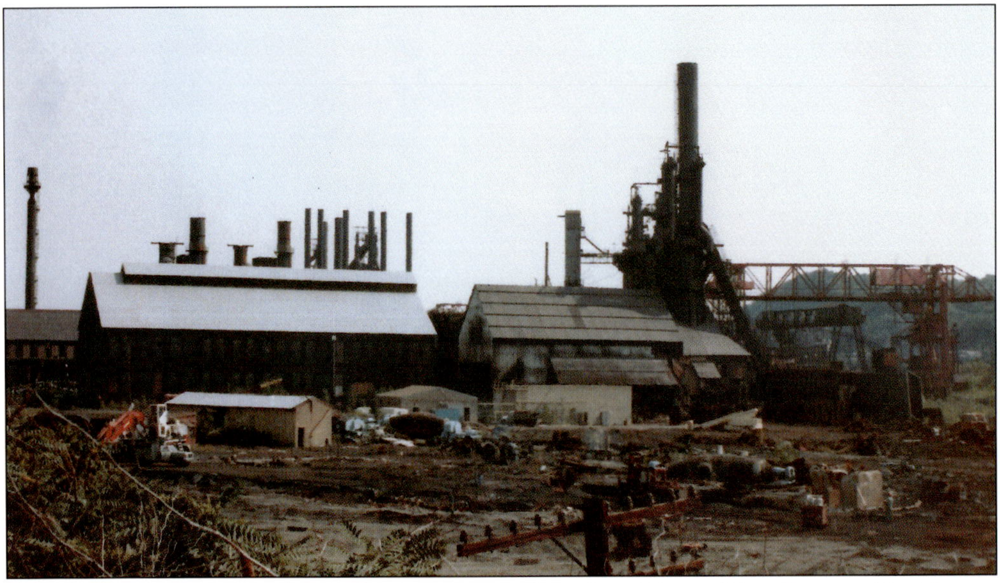

While looking east and a bit south, one can see the side of Carrie Blast Furnace No. 7's cast house and the broken trestle high line for the stock house in 1989. The flare tower and venting stacks still stand. Note the two bottle cars. There is also a man in the lower right, with a red hat and green shirt.

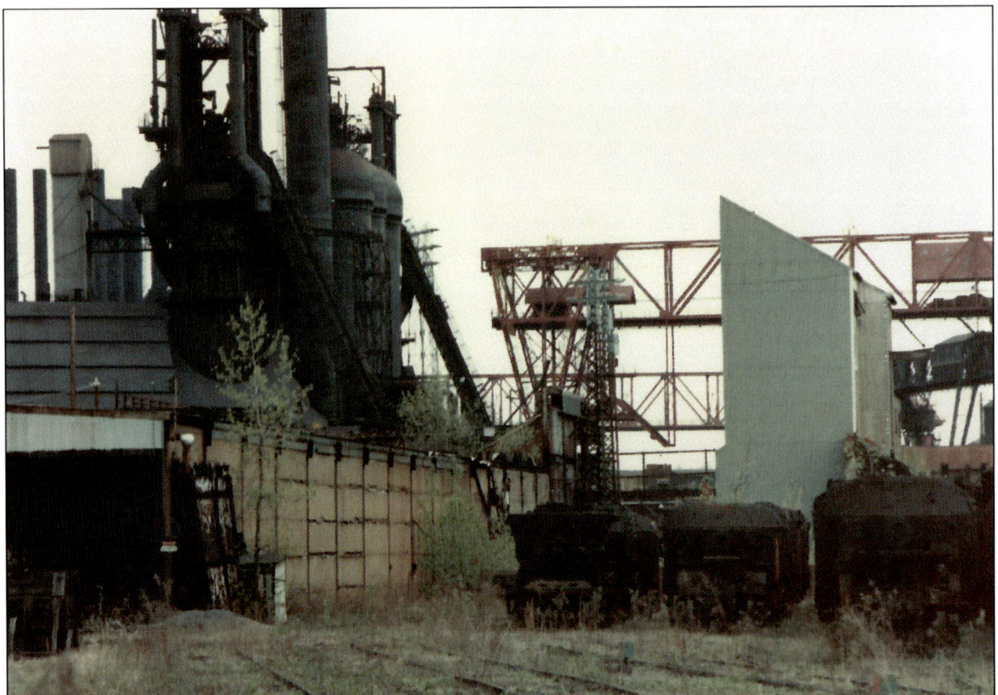

Looking southeast is Carrie Blast Furnace No. 7 with ladle cars on the right side of the photograph. As one can see, the end of the stock house trestle has not been taken off yet and the cast house is intact in January 1988.

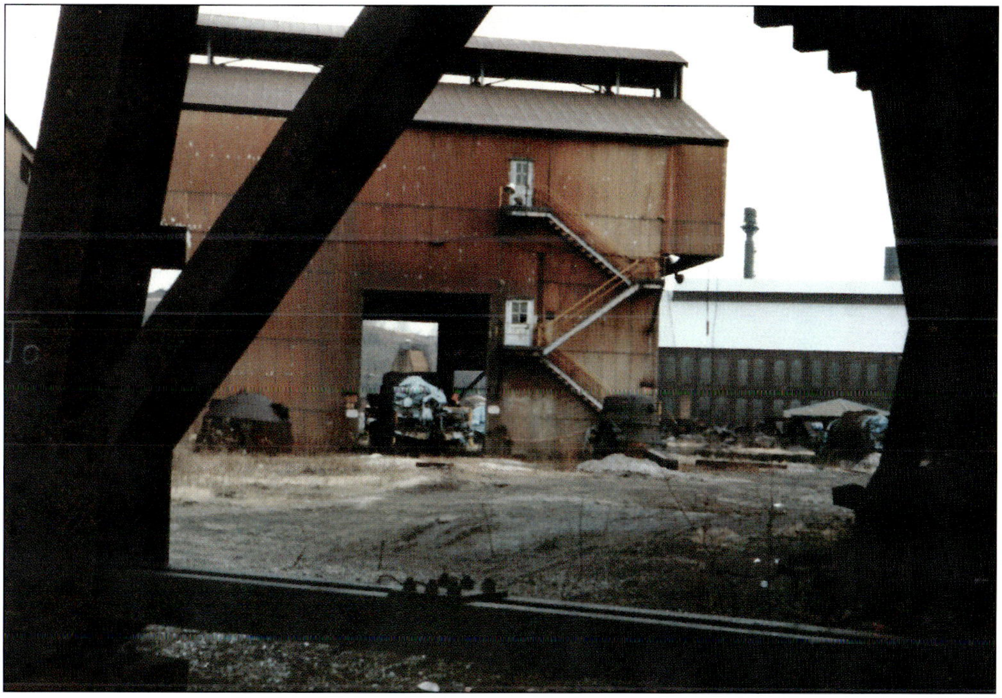

In January 1988, this photograph was taken from under the stocking trestle while looking east at the ladle house, which is no longer there. A bottle car (with a blue end) sits just inside the building.

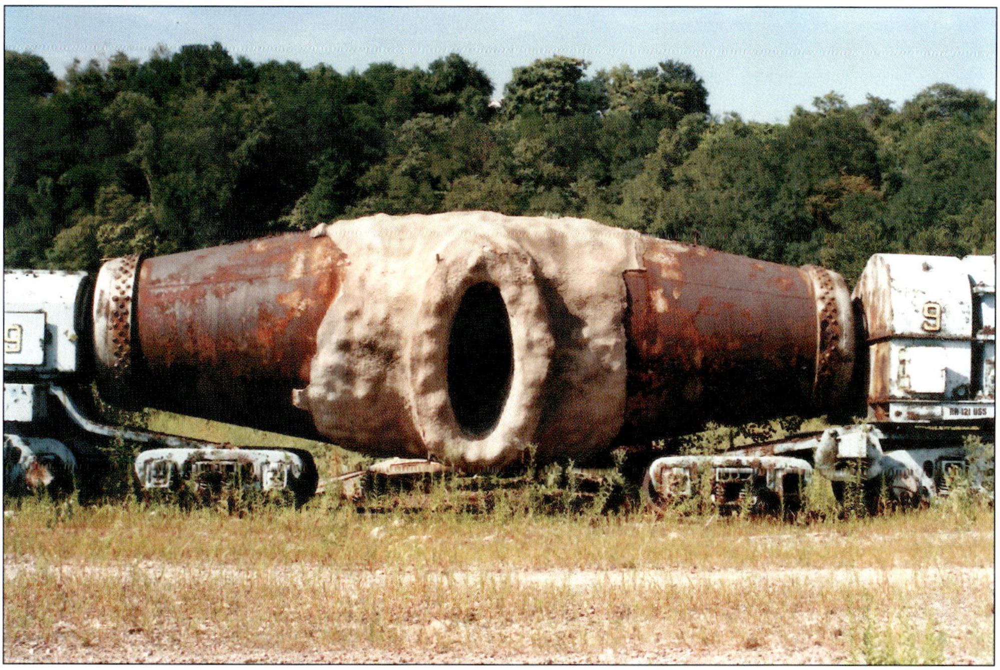

This is the same bottle car that is seen in the previous photograph. Car No. 9 was manufactured by Pollock, and the car type was also called "torpedo," "cigar," or "submarine." As of May 2015, the car still sits on the northwest side of the plant but will soon be moving to Cast House No. 6.

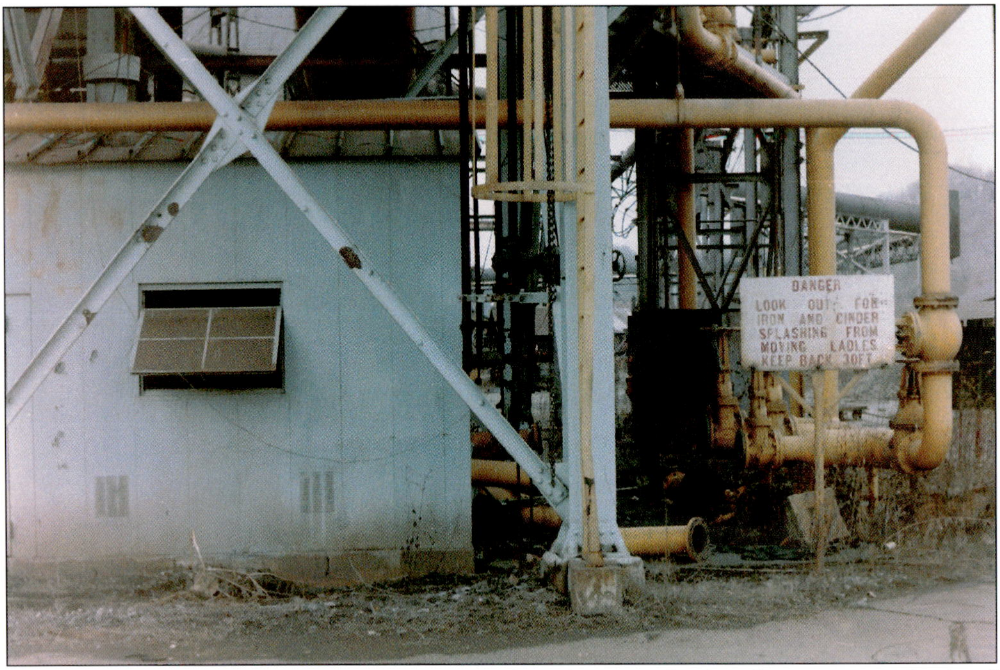

Walking up from the employee entrance gate and heading west for the Blast Furnaces No. 6 and 7, one will find this sign "Danger Look out for iron and cinder splashing from moving ladles keep back 30 feet."

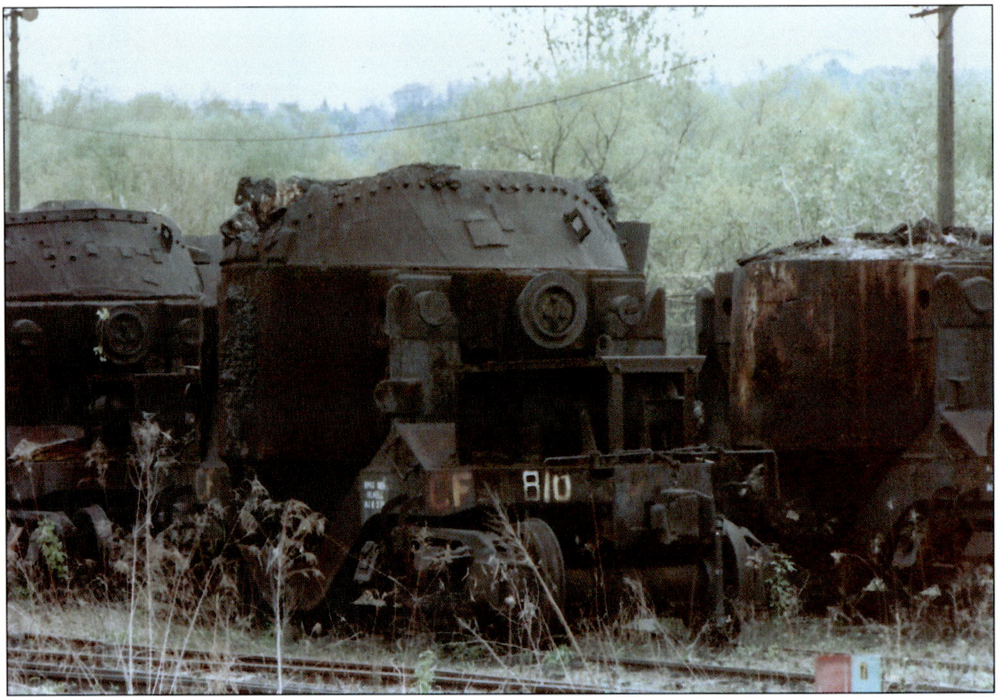

In the southwest yards, these slag pots were lined up waiting to be scraped. Note they were marked "CF." Some were filled with solidified slag, some were empty, and all were rusty. Today, all are gone.

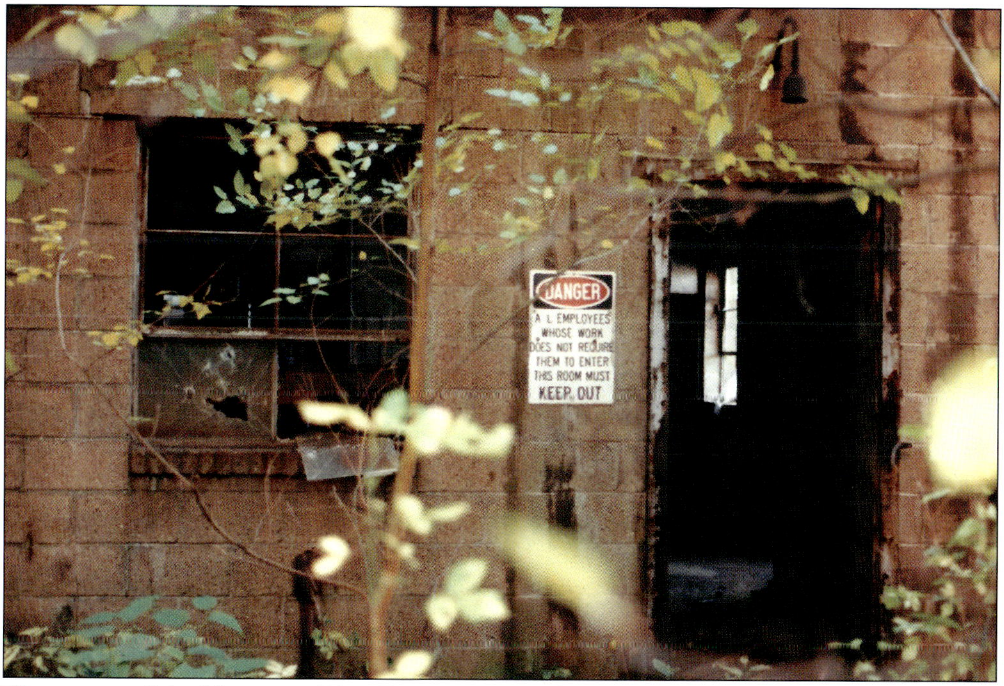

This is a small building near Blast Furnace No. 7; the sign outside the door reads as follows: "Danger—All employees whose work does not require them to enter this room must KEEP OUT." Some pump-like machinery was found inside the small building at the time of the photograph.

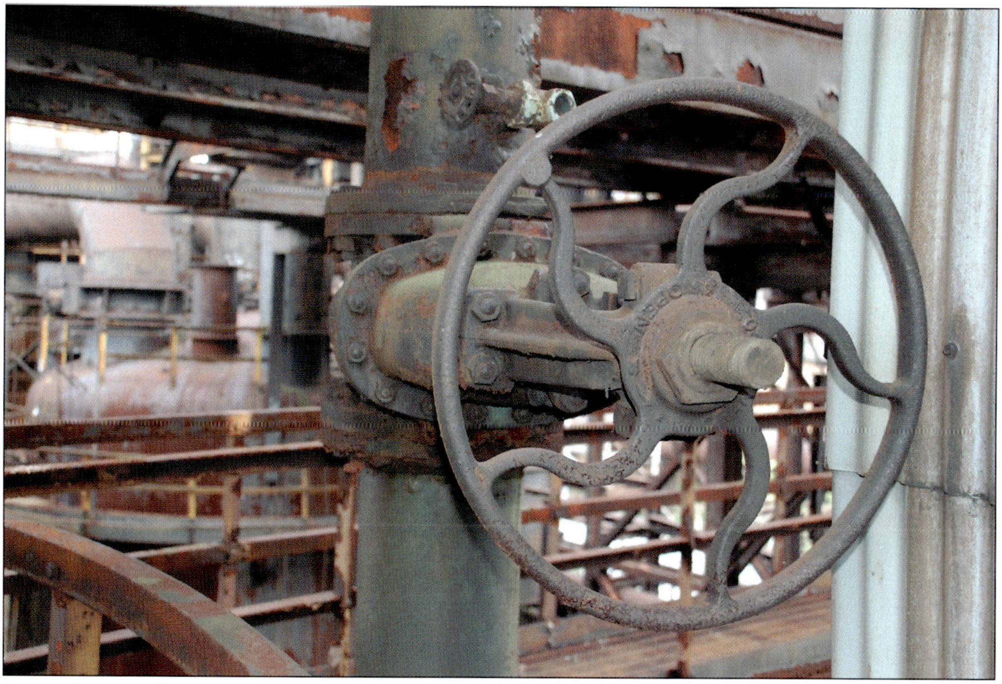

Lots of manually operated valves, pulleys, and chains could be found all over the place.

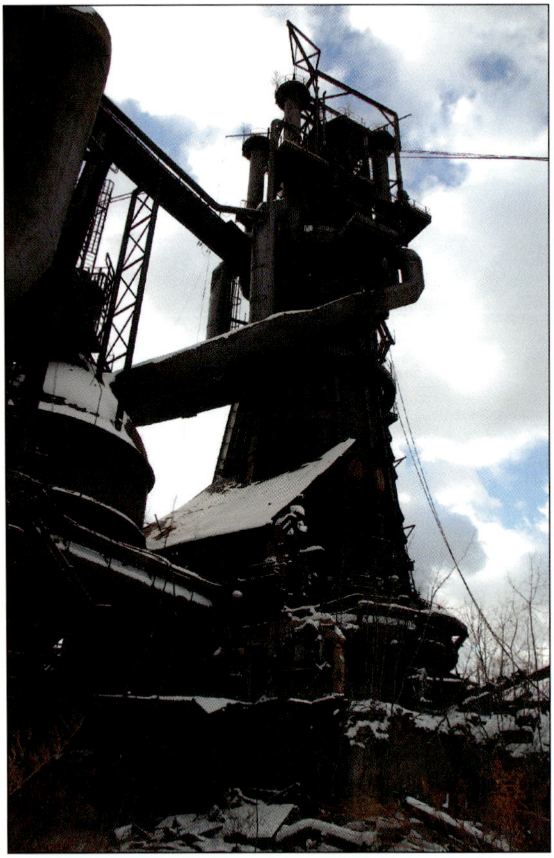

Blast Furnace No. 7 is pictured. Its cast house has been torn off and left open to the elements for 30 years. Built between 1906 and 1907, it stands 92 feet tall, covered in 2.5-inch-thick steel plate and lined with refractory brick to withstand the 3,500-degree Fahrenheit heat. In 1978, it was producing 2,500 tons of molten iron per day.

Here is Blast Furnace No. 7. Looking up into a deep-blue sky, the white clouds almost look like smoke from the furnace.

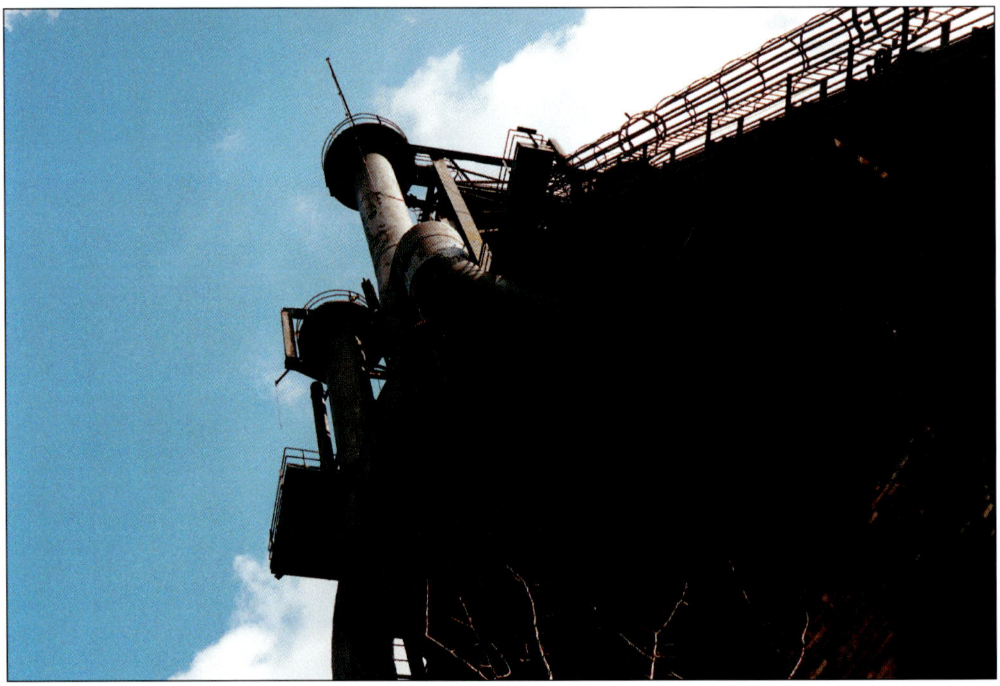

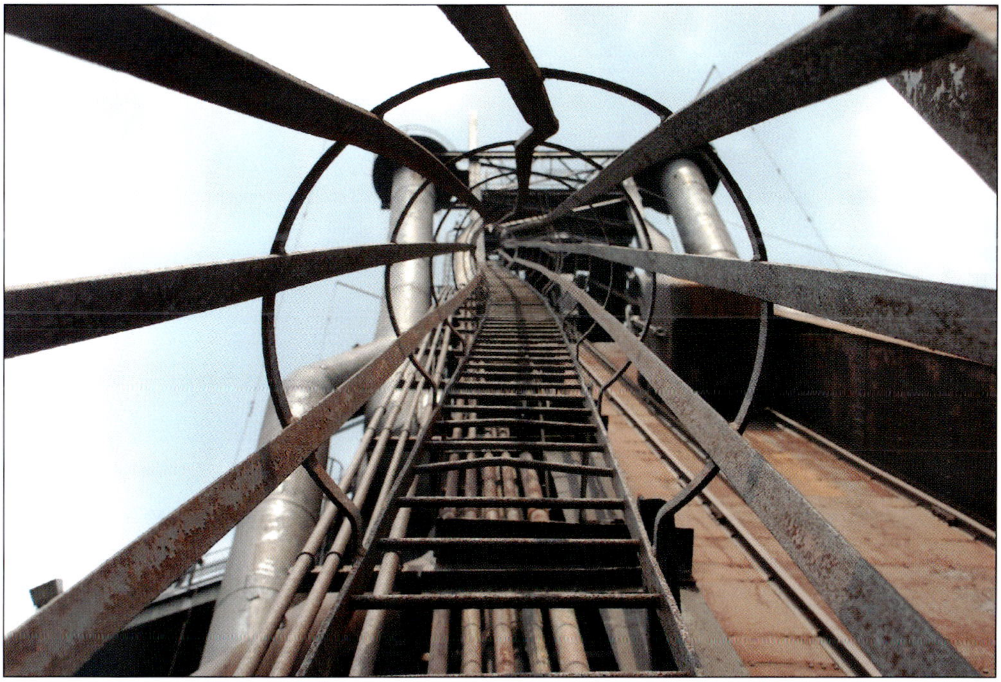

This photograph was taken going up the access ladder alongside the skip hoist (on the right) for Blast Furnace No. 7. The gas flows up the offtake pipe on the left.

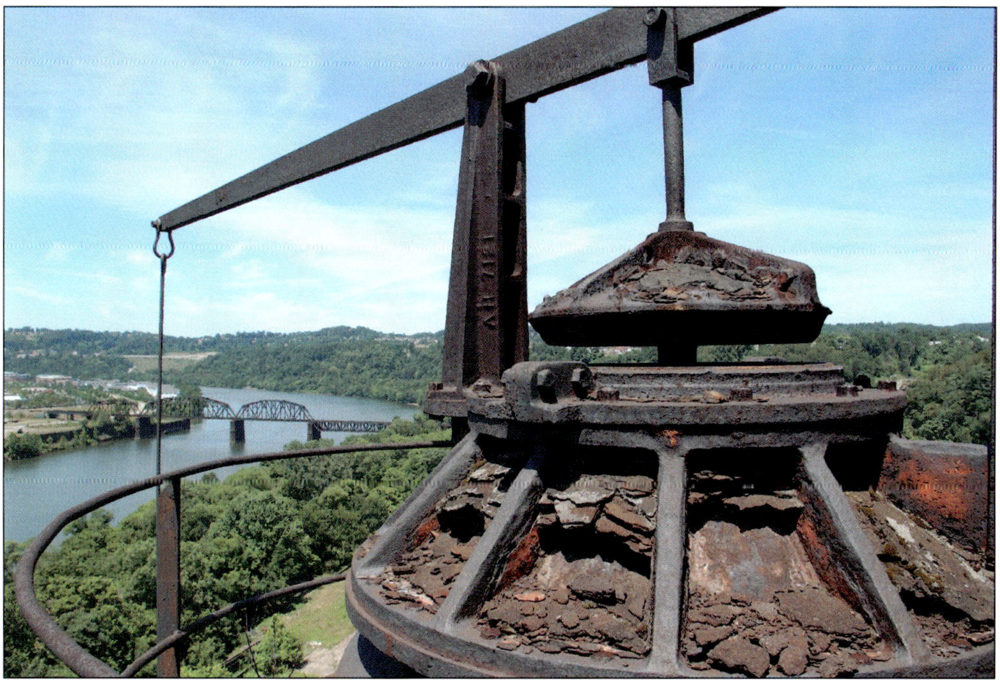

The view downriver (west) shows a part of the mechanism for opening the top of the uptakes, the "bleeder valves," which release gas and protect the top of the furnace from sudden gas pressure surges. Fire and smoke would shoot out of these when the furnace was working.

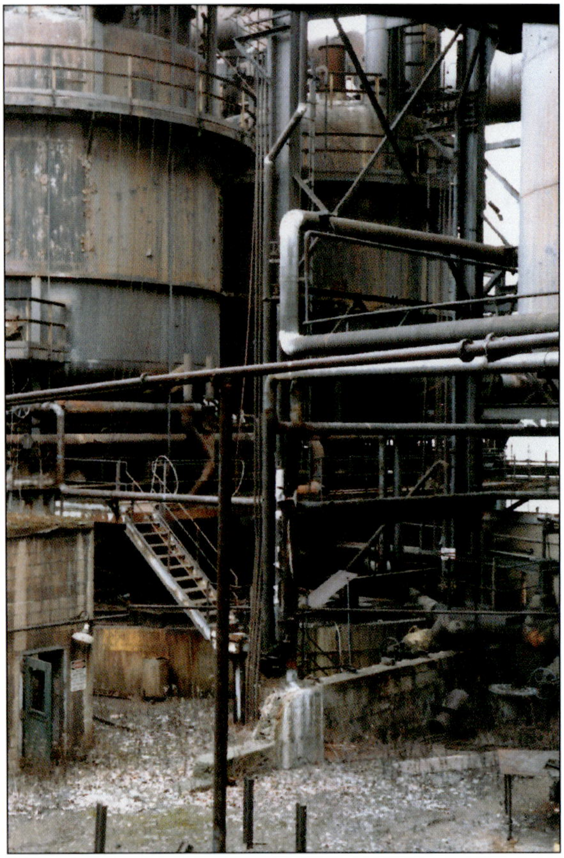

This photograph was taken looking northeast at the electrostatic precipitators at Carrie Furnace. The electrostatic precipitators are a filtration device that removes fine particles, like dust and smoke, from a flowing gas using an induced electrostatic charge, minimally impeding the flow of gases through the unit. This photograph was taken in January 1988.

This photograph was taken from a stairway, looking north at the electrostatic precipitators at Carrie Furnace, which were installed in 1956. There is wonderful symmetry built into these, as well as wonderful colors.

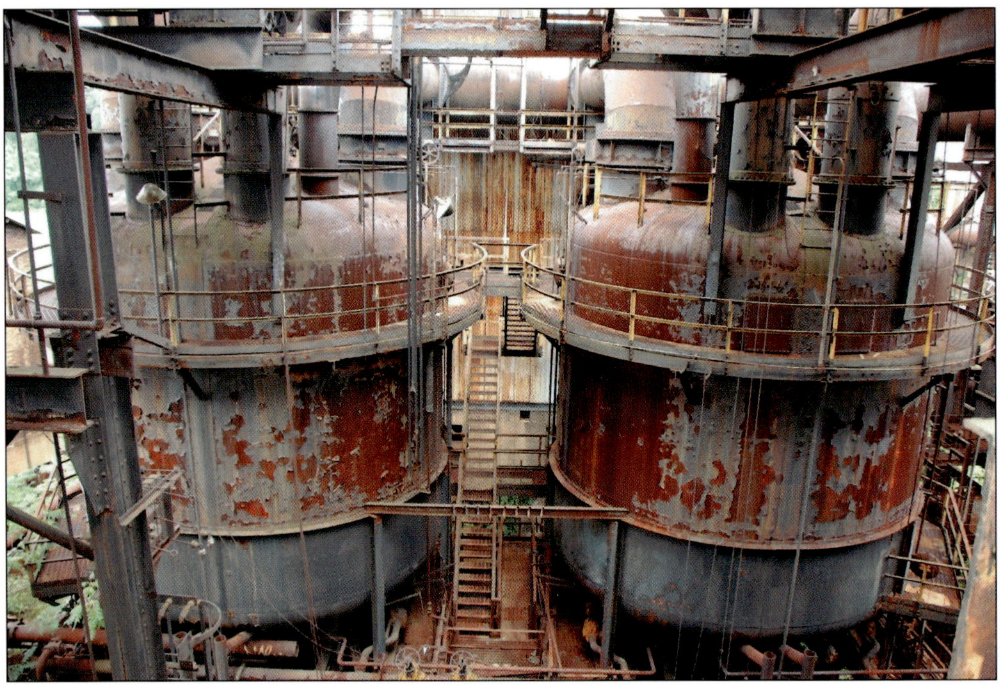

This was photographed on top of the electrostatic precipitators, which were installed in 1956. Looks like two outlets of some kind, operated by chains from below, were on top of the units.

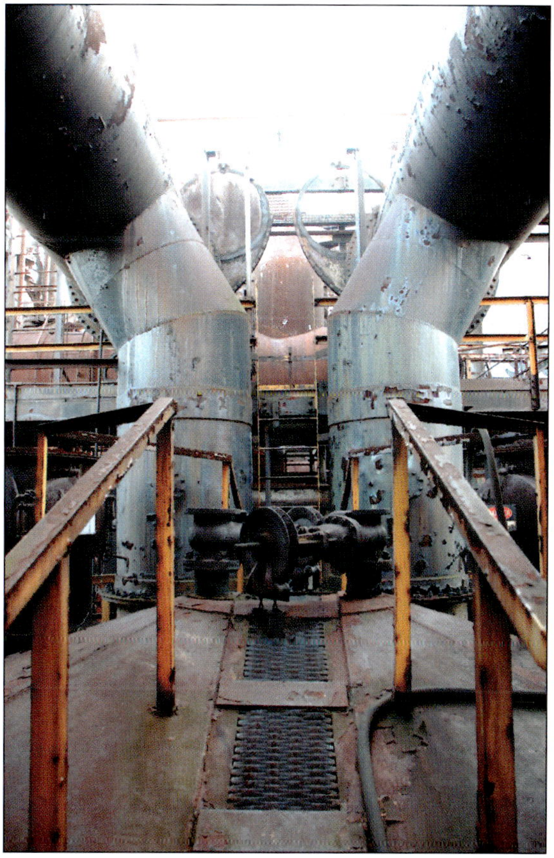

This photograph is titled "Sanctum Santorum." It must have been a sight if, after walking up the stairs to the top of these giant cylinders, one mounted the last few steps into the center and were in the middle of all those surrounding pipes in the dark of night. Even in this state, it is awe inspiring to be there.

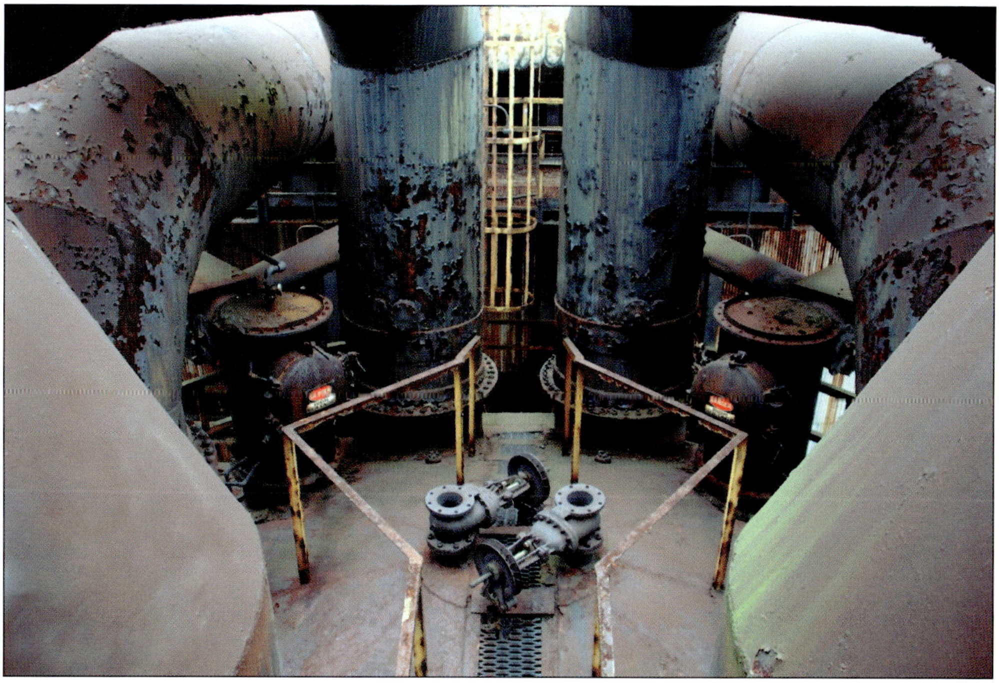

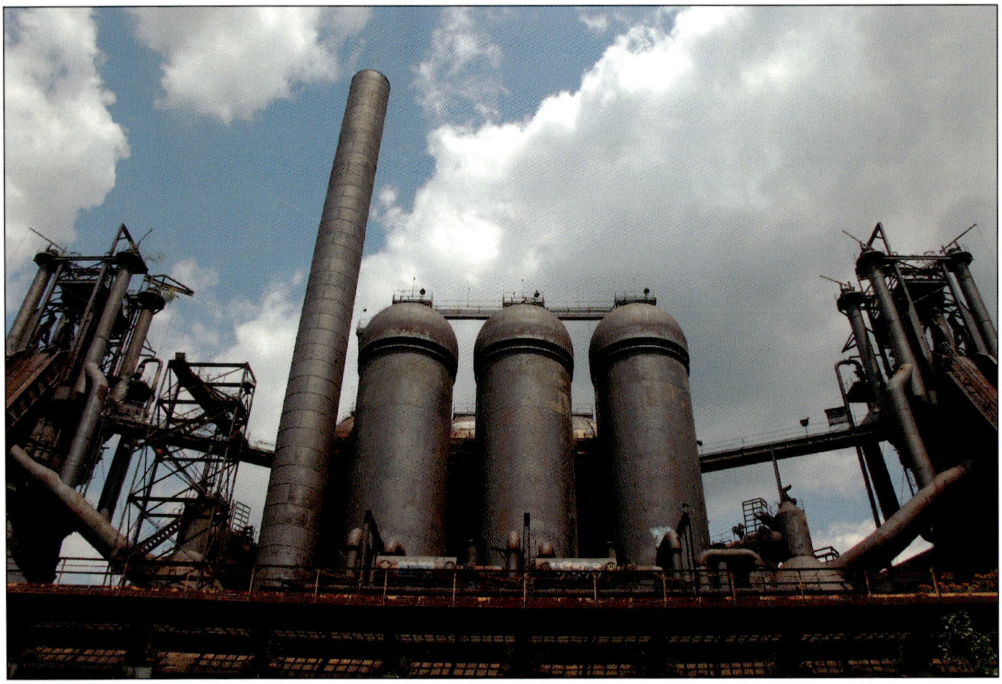

Standing in the ore yard, looking north, one sees, from left to right, Carrie Blast Furnaces Nos. 7 and 6. There are draft stacks on the left and three Cowper hot-blast stoves in the middle. Together with the blue sky and clouds, this makes a striking image.

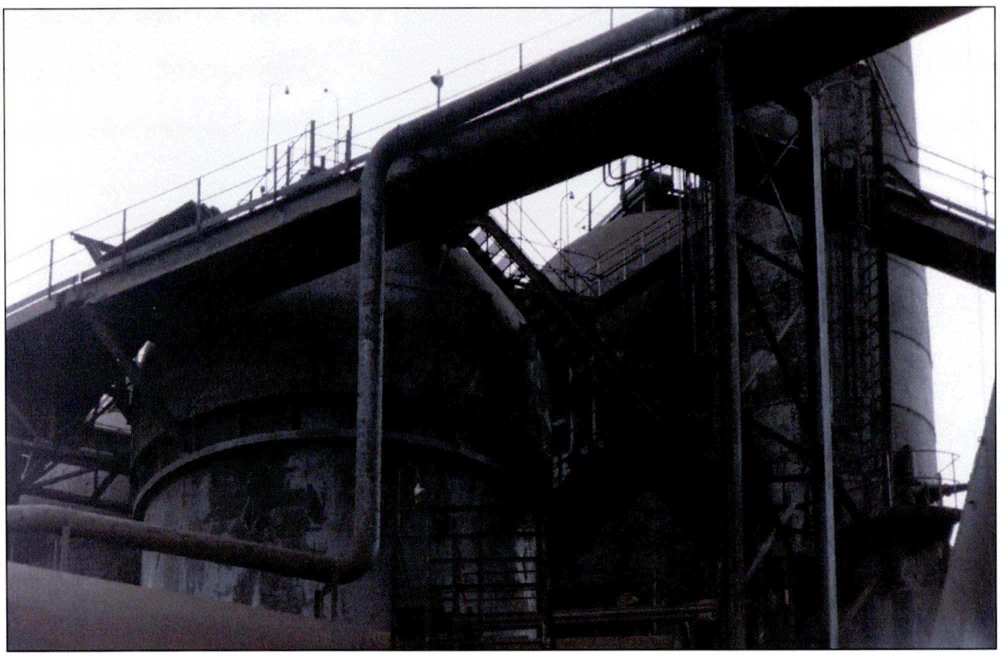

Here is a view looking up at the tops of the Cowper hot-blast stoves from ground level. Also pictured is the walkway behind and between them. These Cowper stoves were installed in 1936.

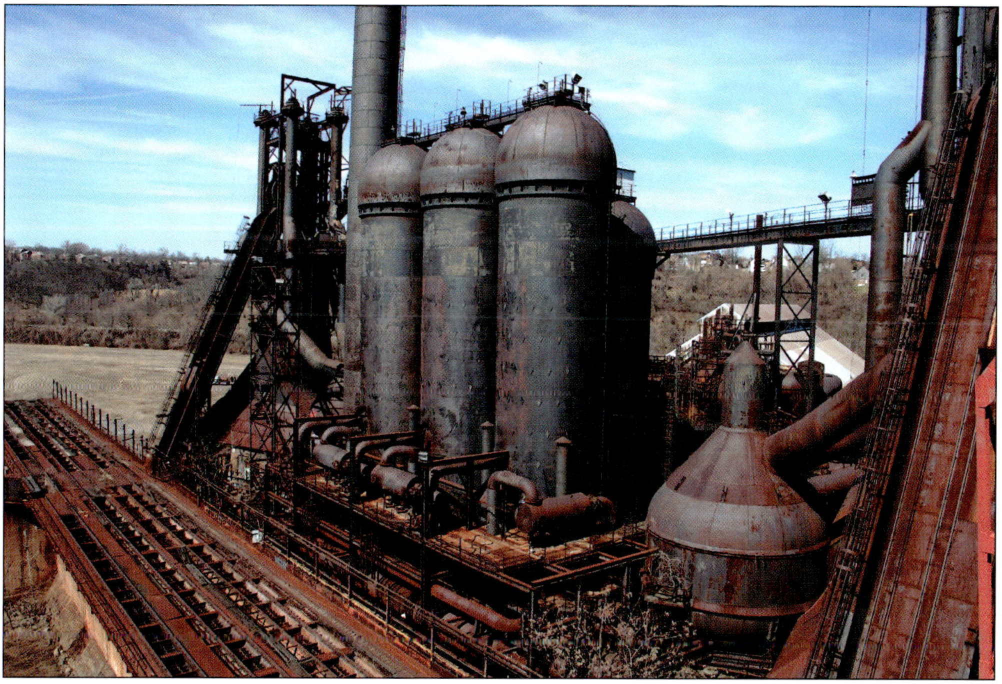

Here is the Carrie Furnace site, as seen from the ore bridge. The high-line trestle runs along the bottom left of the photograph, and Blast Furnace No. 7 with skip hoist is on the far left with the draft stack and Cowper hot-blast stoves on the right. The dust catcher for Blast Furnace No. 6 is seen beneath the catwalk.

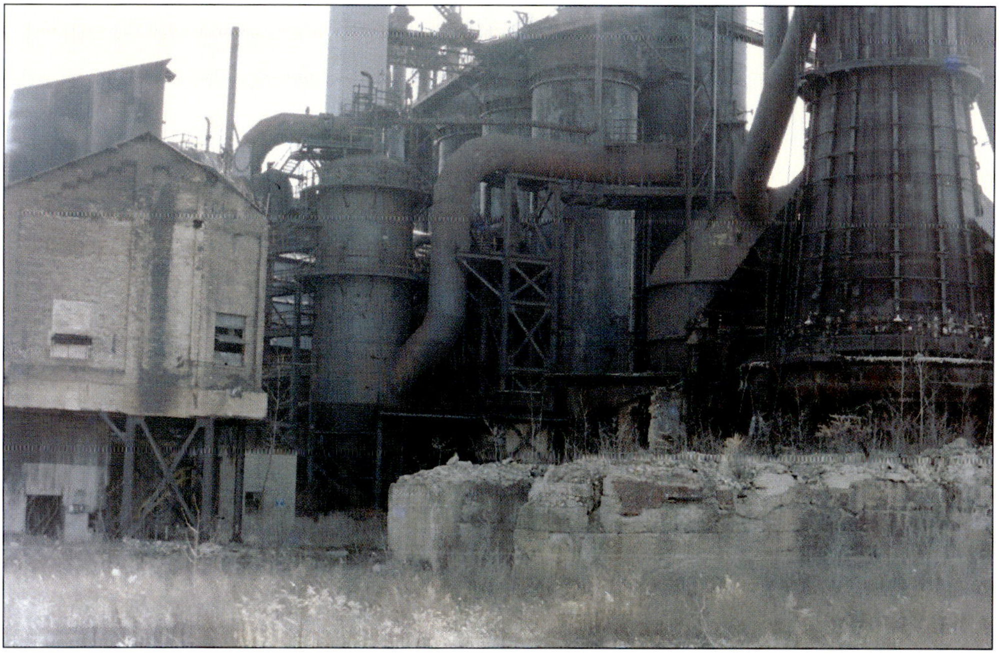

From the other side of the blast furnace, one can see, from left to right, the hoist house for Blast Furnace No. 7, gas washer (installed in 1968), Cowper hot-blast stoves, a dust catcher, and a destroyed cast house.

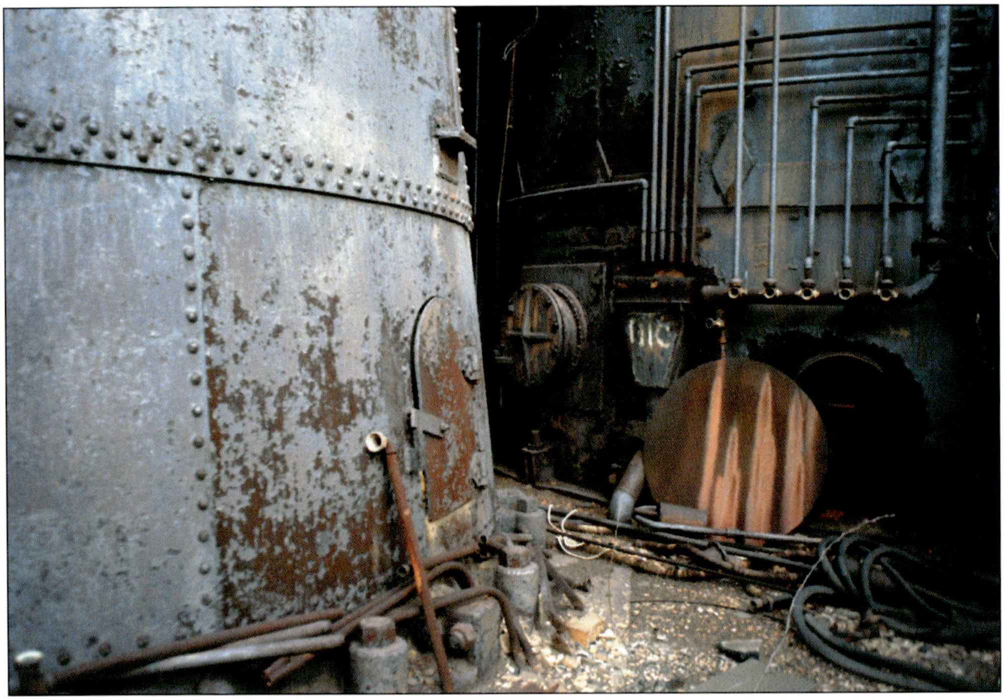

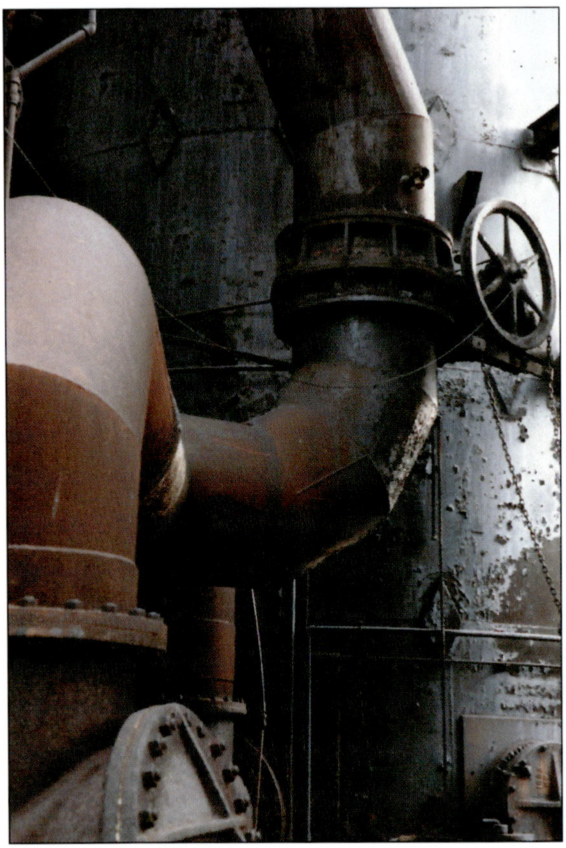

On the left is the base of the draft stack. Note the little door on the side. Some cable is lying on the ground, and the rain has made interesting rust streaks on that metal disc.

This January 1988 photograph shows the interesting ductwork at Carrie Furnace in the area of the Cowper stoves. There were not a lot of graffiti tags all over the place at the time of this photograph.

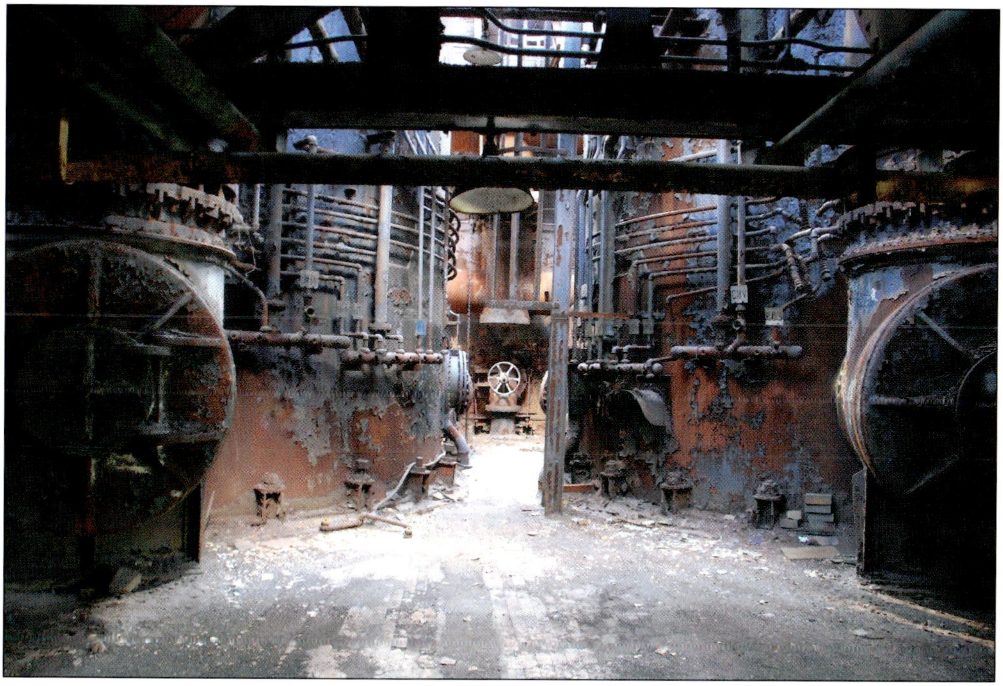

Again, the symmetry of these Cowper hot-blast stoves is remarkable. This photograph was taken in 2009, and the stoves were a lot rustier than when they were photographed 20 years before.

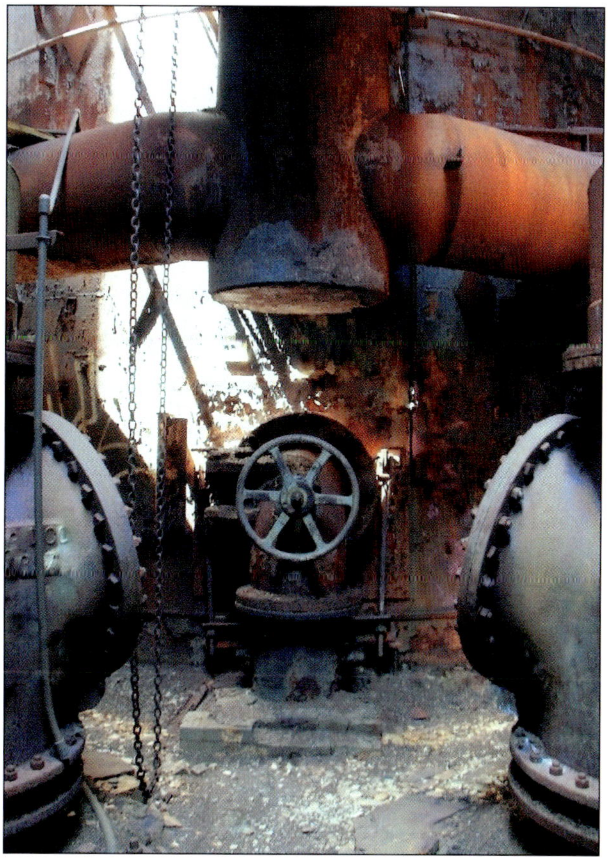

Among the Cowper hot-blast stoves, one can see valve handles, hanging chains, giant pipes, peeling paint, and the red rust of metal and can hear the lonely sound of something high up in the furnace banging rhythmically with the wind.

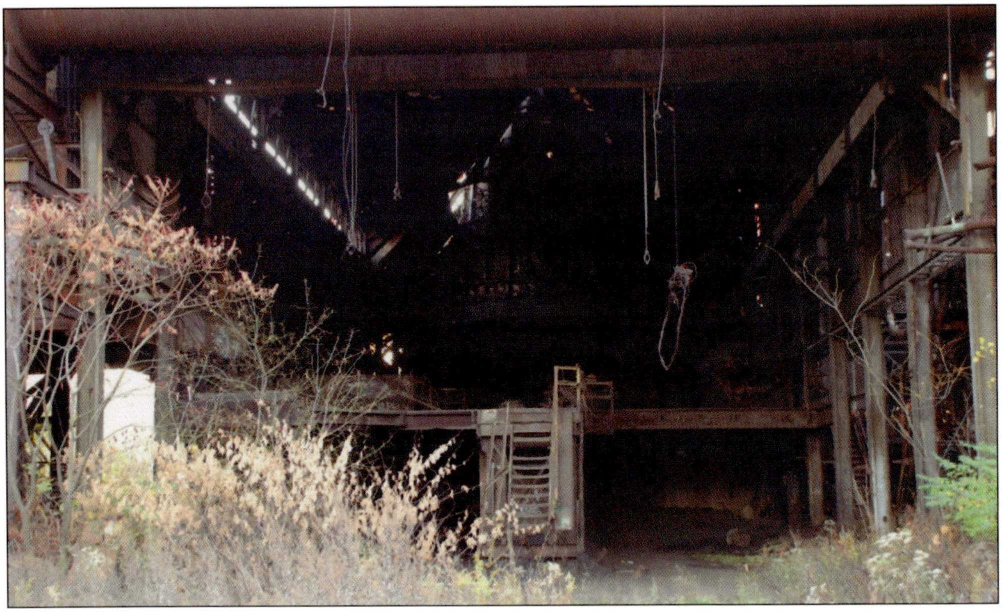

Approaching Carrie Blast Furnace No. 6., one can see into the cast house from ground level. The bottom of the stairs have been cut off to make it a little more difficult for people to get into the plant, something that was done throughout the site as a deterrent.

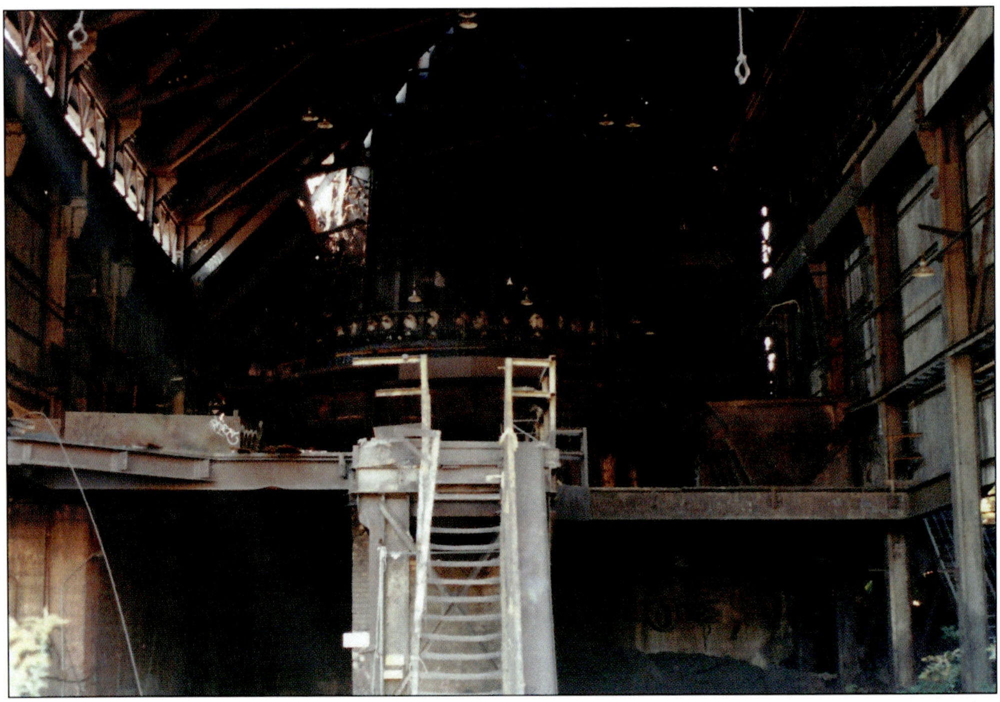

Getting a little closer, the cold empty Blast Furnace No. 6 sits quietly and majestically in the stillness of the morning. The sun's morning rays are shining in from the left. Some slag remains under the furnace.

This October 2009 view shows the central runner with diverters down to channel the molten iron/slag off to the sides. The ring around the bottom of the furnace is the bustle pipe. The tap hole is in the center at the bottom of the furnace. The control room is just off to the right. On the left are bags of clay used for plugging the tap hole.

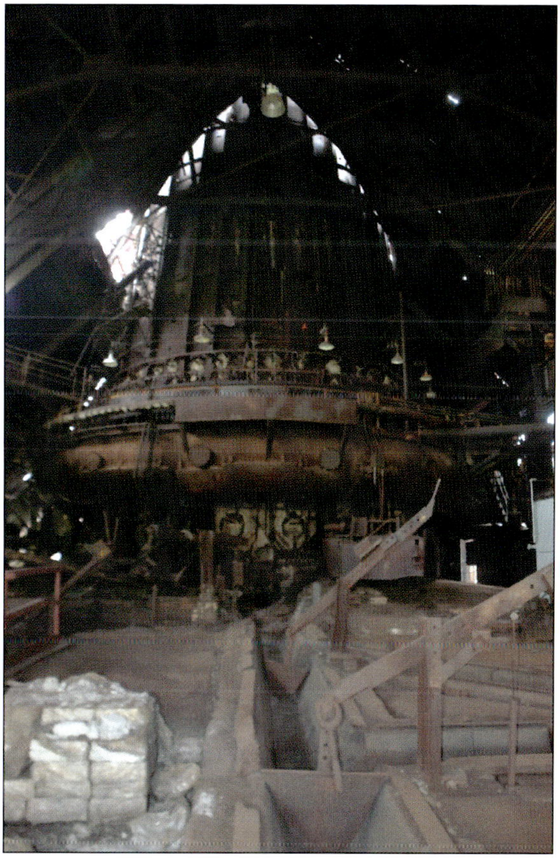

Here is another view of Blast Furnace No. 6. Wires from the control mechanism on the cast house floor (at right) went to the diverter mechanism by the runner to divert the flow of molten iron and slag from the furnace.

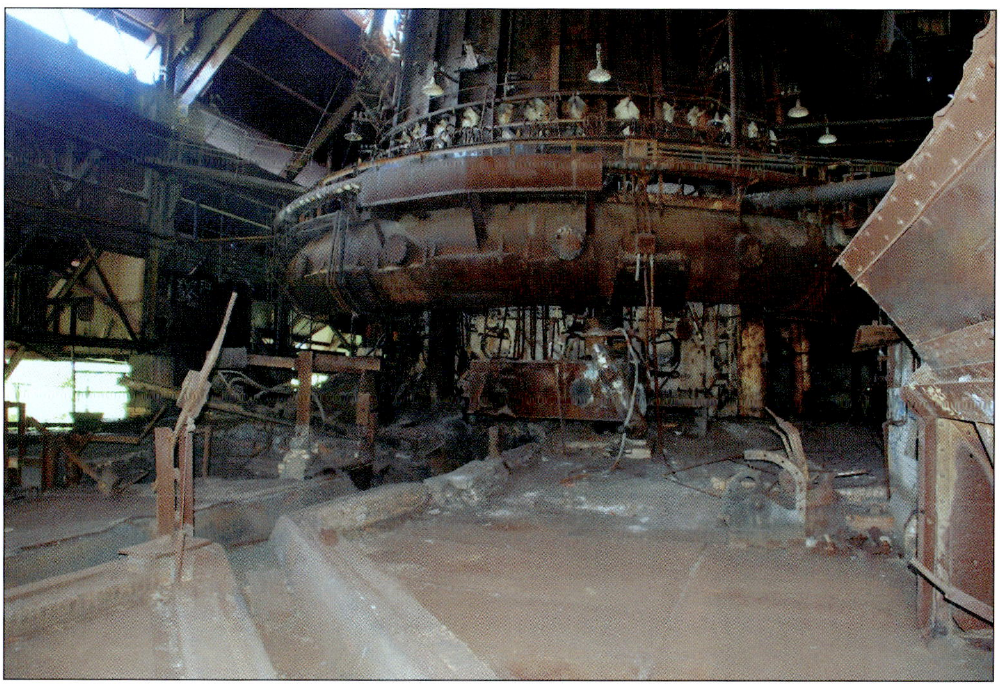

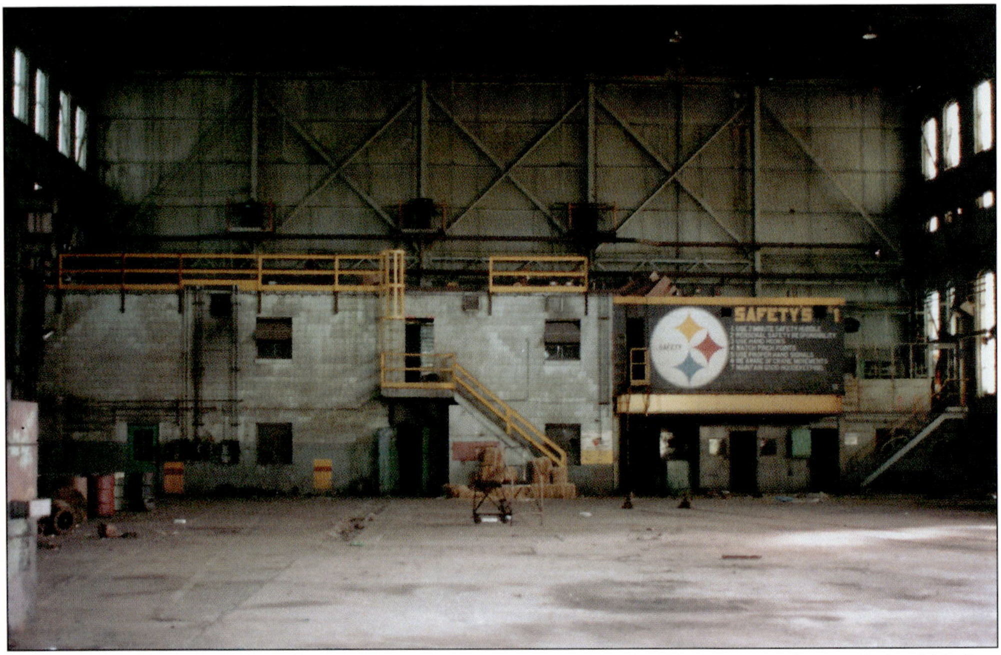

Here is a look inside the emptied blowing engine room. The building is 104 feet wide by 300 feet long with a 50-ton overhead crane. It was built in 1907 and refitted in 1956, and it was connected to the cold blast main. Note the steel mark painted on the back wall; it belonged to the American Iron and Steel Institute (AISI). It is a circle inclosing three hypocycloids, which is also used by the Pittsburgh Steelers football team.

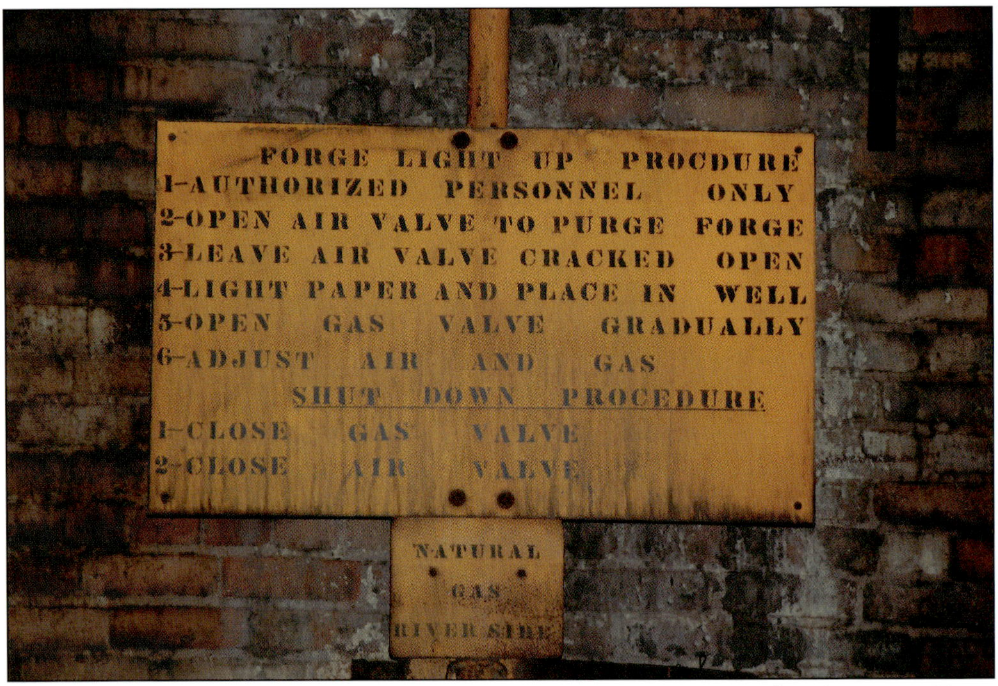

The following was found along the east wall of the blowing engine room, and reads, in part: "Forge light up procdure (sic)." It is one of the few signs not vandalized or covered with graffiti.

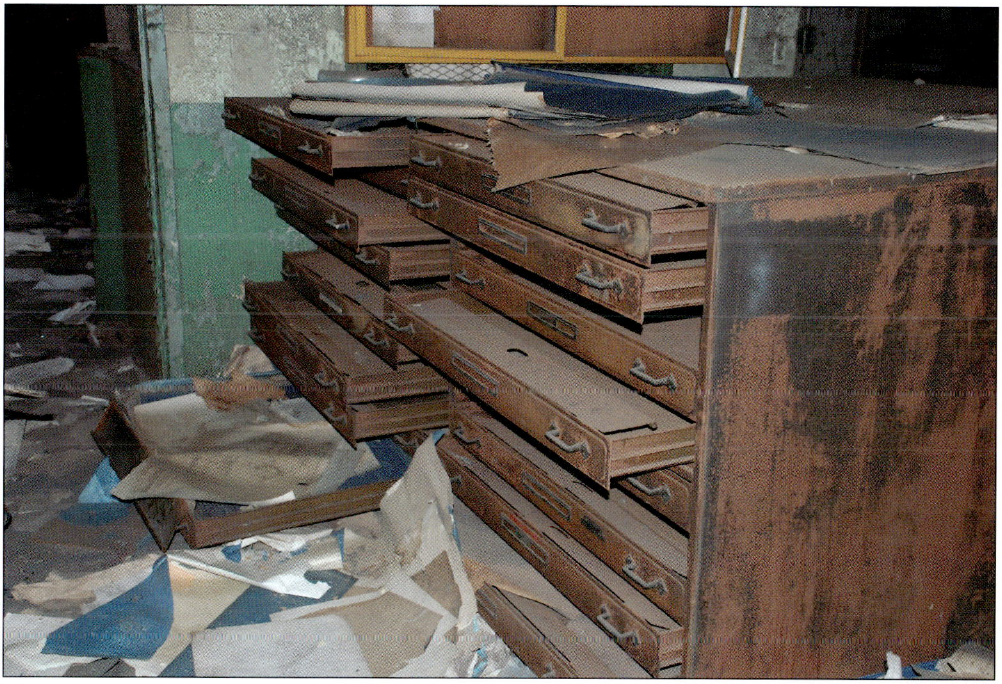

The office in the back end of the blowing engine room remarkably, even 25 years after closing the area, still had drawings and other papers strewn about the floor. There are even logbooks of who worked where and when.

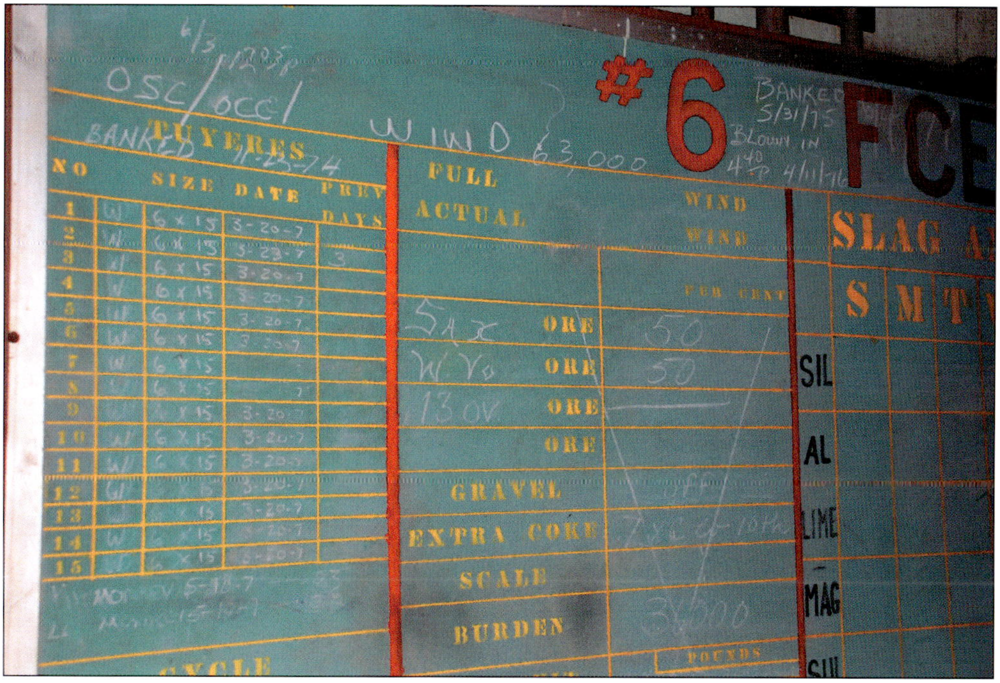

Part of one of the control rooms behind the Cowper hot-blast stoves, separated from the chart recorder room, had two of these green boards with the following information chalked on them: "#6 FCE banked 5/3/75" and "Blown in 4:40 P4/11/76."

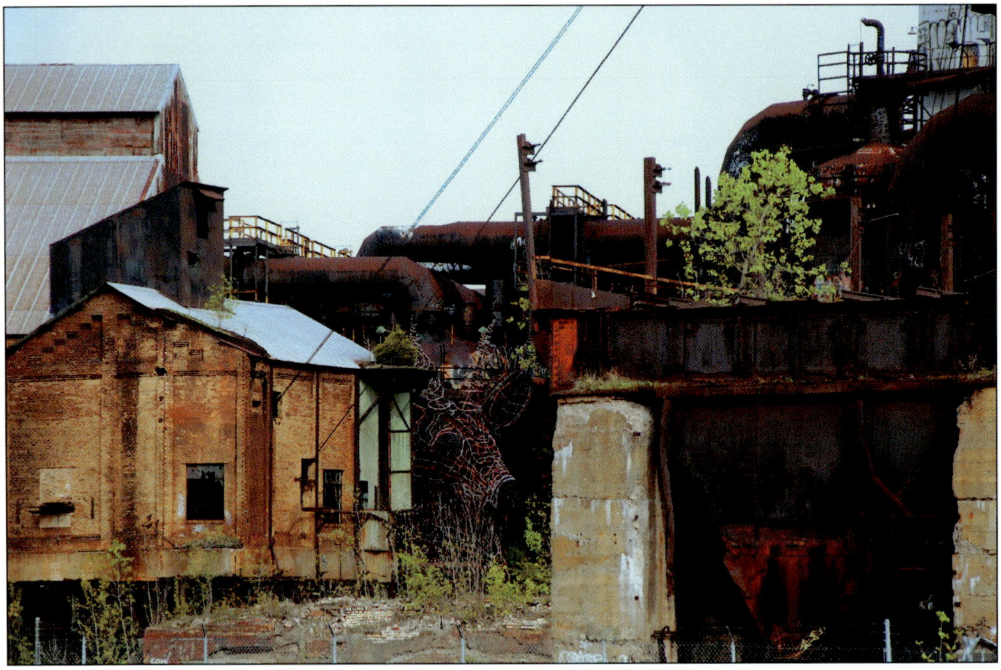

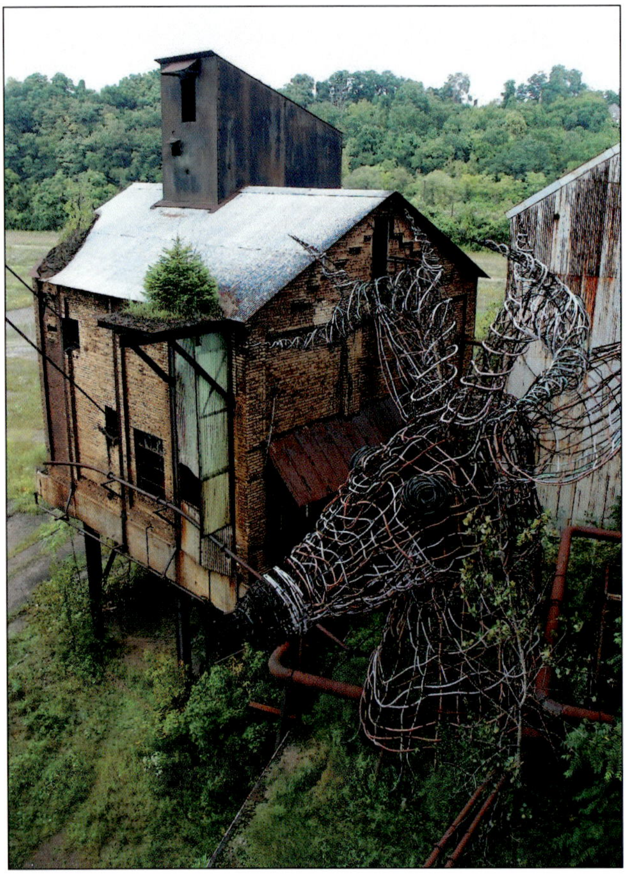

The 45-by-35-foot "Rankin Deer" is an installation sculpture built on site using materials found there. It was created by six or eight artists, from the industrial arts cooperative, working Sundays between October 1997 and October 1998. The "Rankin Deer" is peeking out from behind Hoist House No. 7. It has been adopted by the Rivers of Steel Heritage group and is being maintained and preserved.

Here is another view of "Rankin Deer" with Hoist House No. 7 behind it on the left.

Seen here is the last shadow man in the plant. He would come out in the evenings when no one was there. A colorful type but quiet. He never said a word, and when one moved, he disappeared.

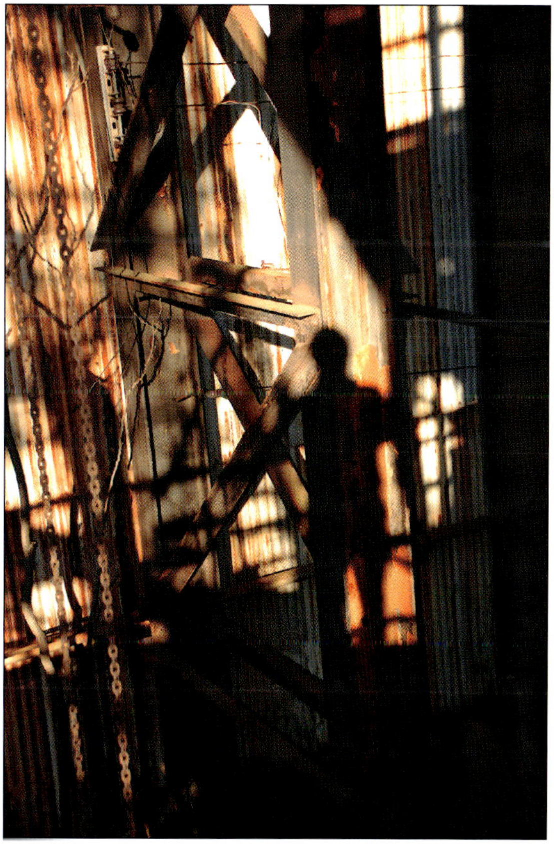

This is a view of Carrie Furnace's hot metal bridge from downstream. It was built by the Keystone Bridge Works in 1900. Note its reflection in the still waters of the Monongahela River. Rankin is on the left, and Whitaker is on the right side.

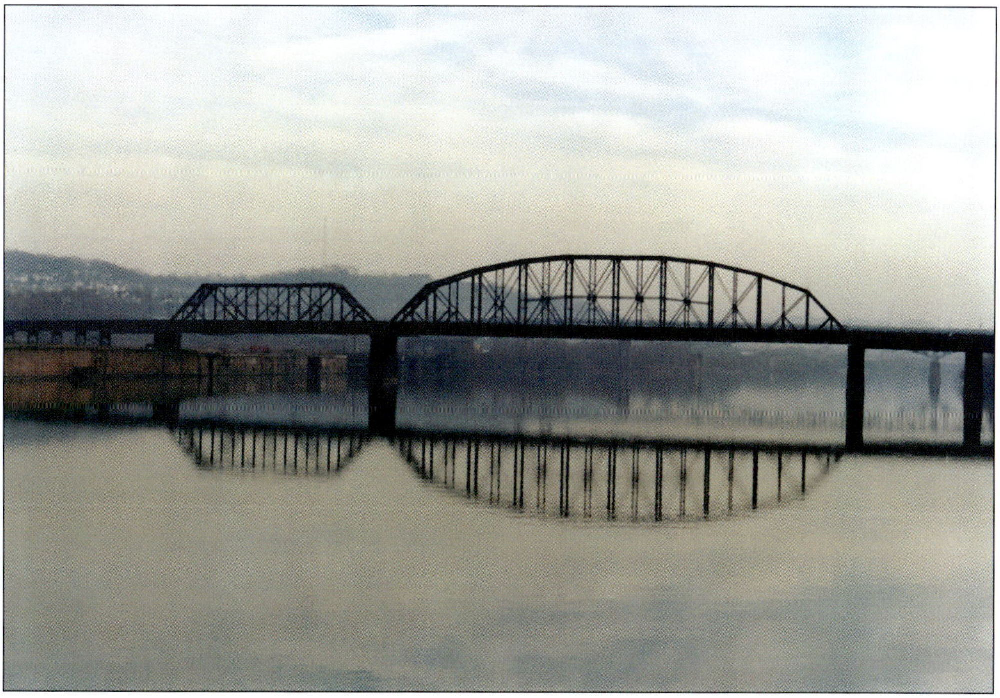

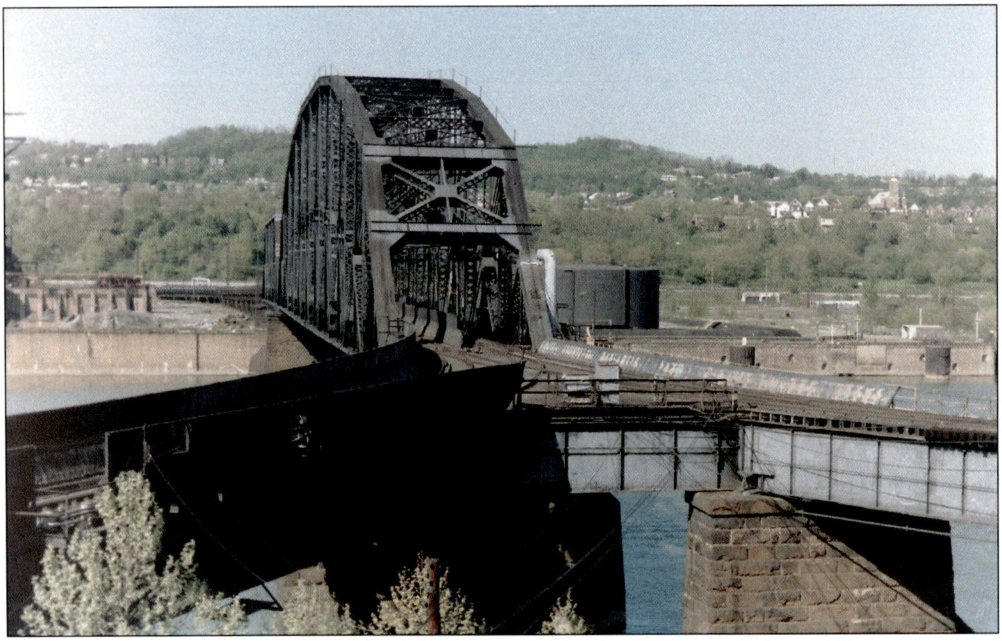

The south end (the Whitaker side) of the hot metal bridge is pictured above. The ramp going to the left took molten iron from Carrie Furnace, and the right side took materials over to the furnace.

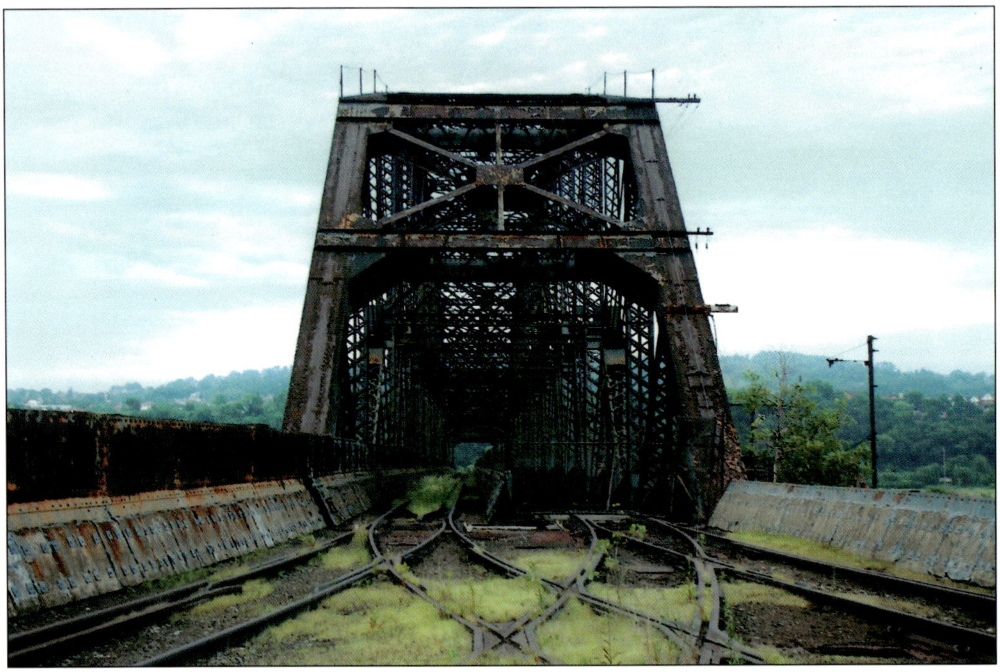

The two bridge builders' plaques on the south end (the Whitaker side) of the hot metal bridge are located on the upstream (right) side and read, "Keystone Bridge Works Pittsburg, 1900," and below that, another plaque reads, "American Bridge Company inspected and reinforced 1915." The bridge took the molten iron from Carrie Furnace over the Monongahela River to the open-hearth furnaces at the Homestead plant.

Five

Jones and Laughlin Plant

Eliza and South Side

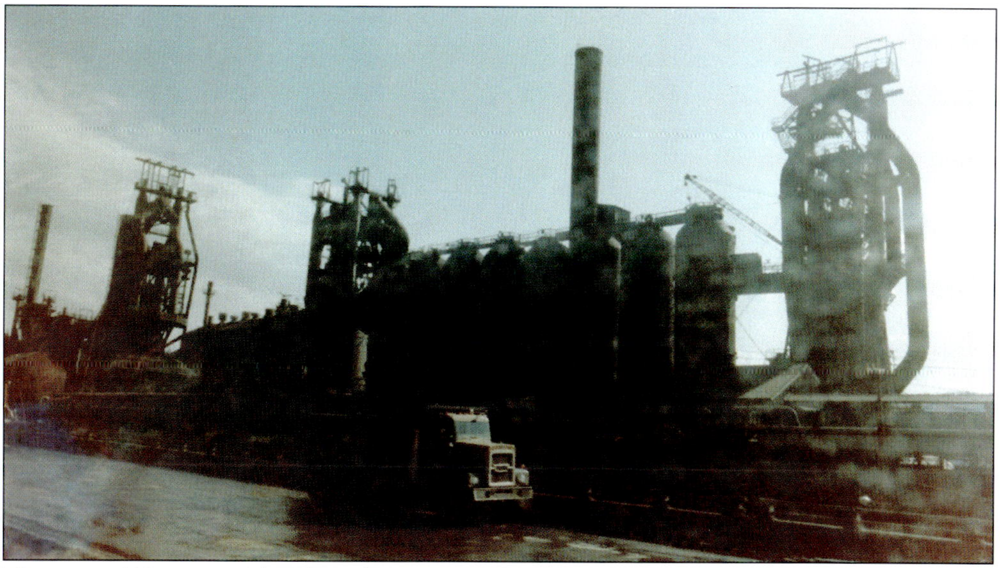

This is a c. 1984 view of the Eliza Blast Furnace from a moving PAT (Port Authority Transit) bus traveling east on the parkway (Interstate 376) in the westbound lanes because the road was being resurfaced and new center dividers and shoulder rails were being installed.

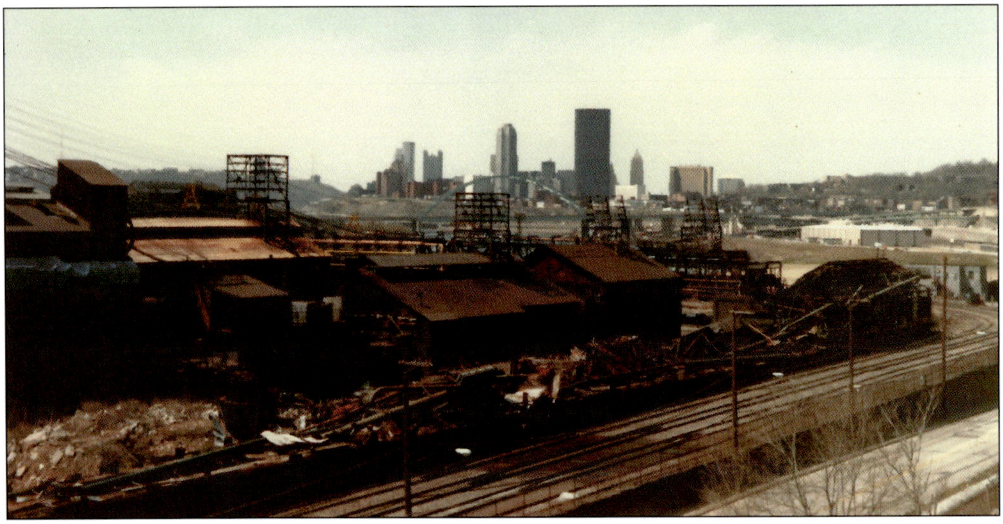

Here, around 1990, is the Eliza Blast Furnace of the Jones and Laughlin site along Second Avenue in Pittsburgh, Pennsylvania. Looking east from a South Oakland hillside, one can see the downtown Pittsburgh skyline. The tallest building is the US Steel tower. Many of the Jones and Laughlin buildings have already been demolished in this photograph. This area is now the site of a technology park.

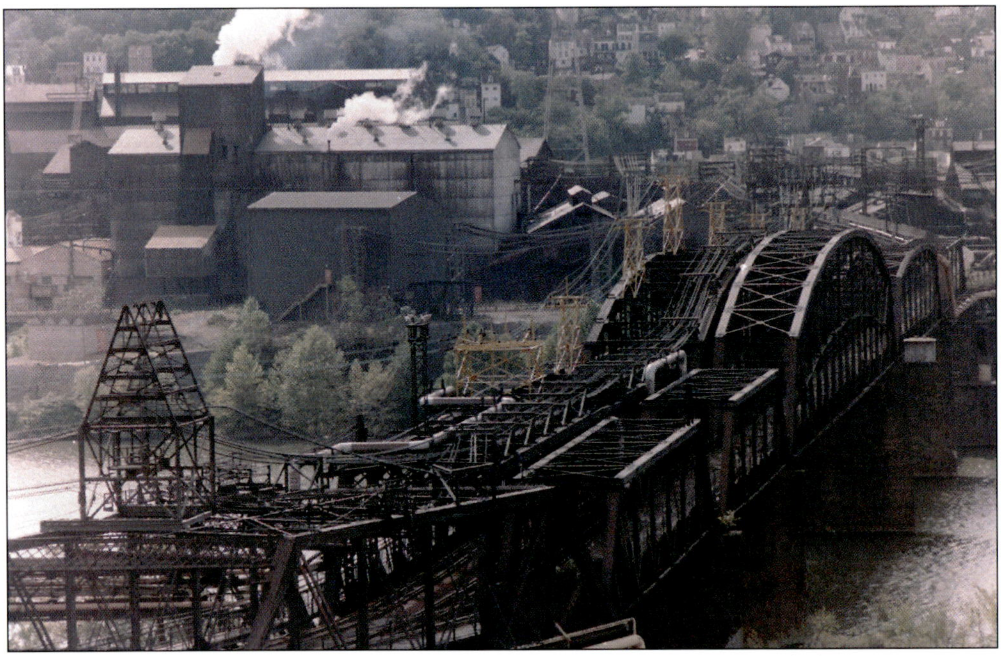

Here, one can see two bridges built on one set of piers over the Monongahela River, crossing between South Oakland's Second Avenue and the South Side's Twenty-Ninth Street. Total length was 1,174 feet. Vertical clearance was 48.4 feet. The railroad bridge on the upstream side was built in 1887 for the Monongahela Connecting Railroad, and the Jones and Laughlin's hot metal bridge was built on the downstream side in 1900. This south-looking photograph was taken around 1985.

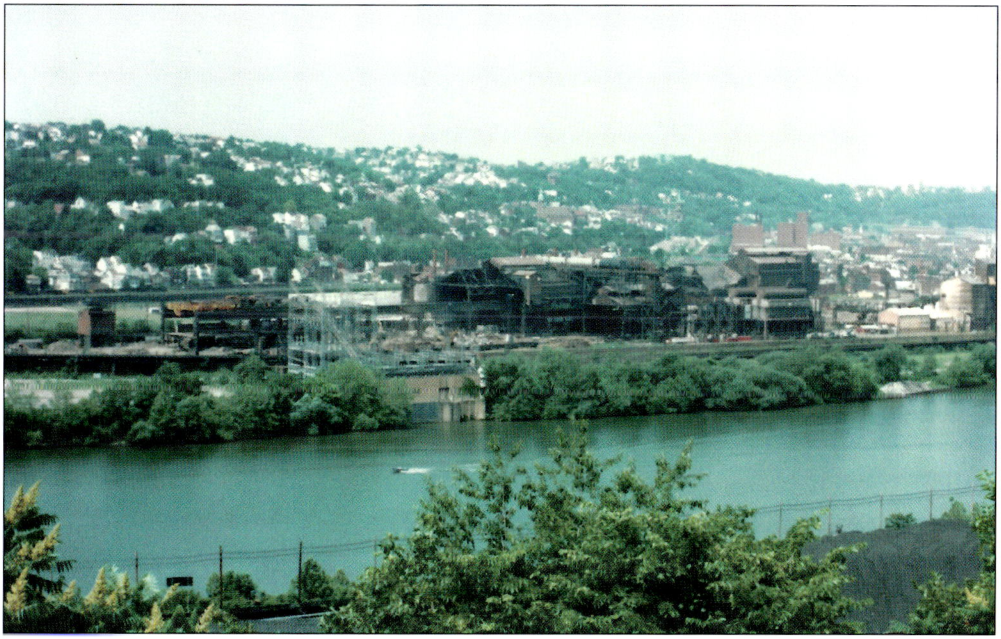

Standing on a hill in South Oakland looking across the Monongahela River, one can see Jones and Laughlin Steel's South Side works as it was being demolished around 1990.

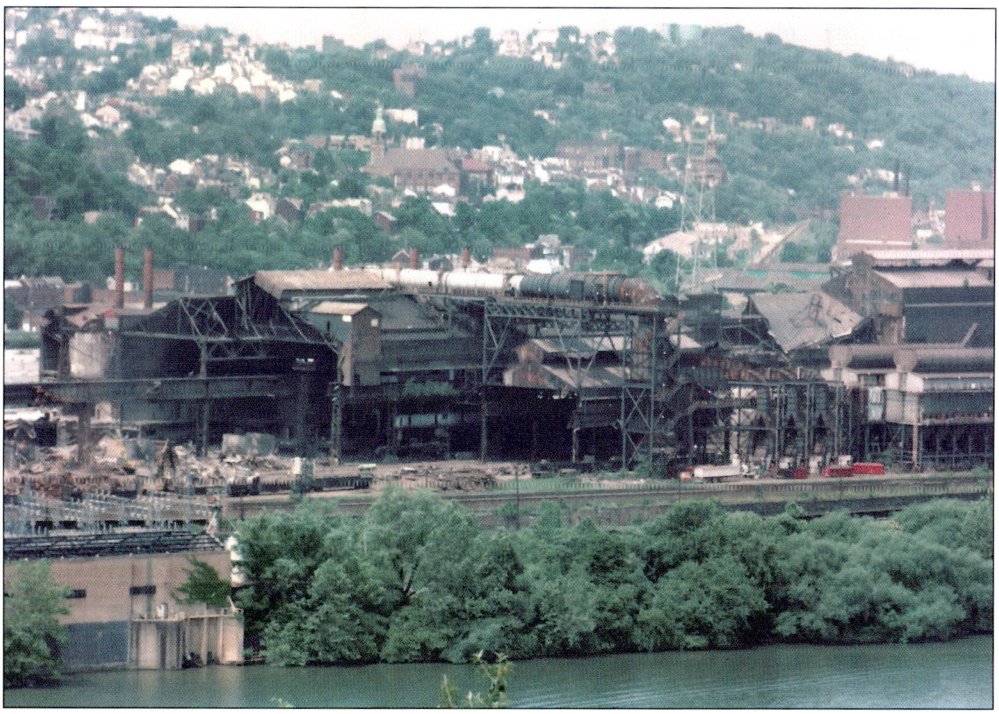

Here is the same general view of the Jones and Laughlin Steel's South Side works being demolished.

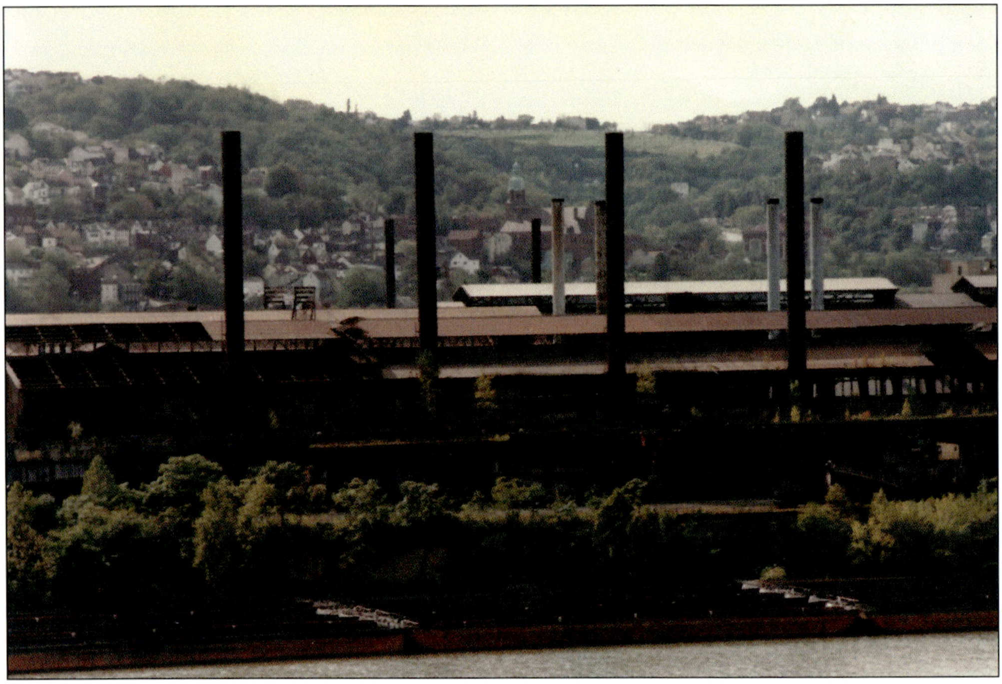

Here are draft stacks of Jones and Laughlin Steel's Open-Hearth Building No. 2, located along the Monongahela River. Today, this area is Riverside Park.

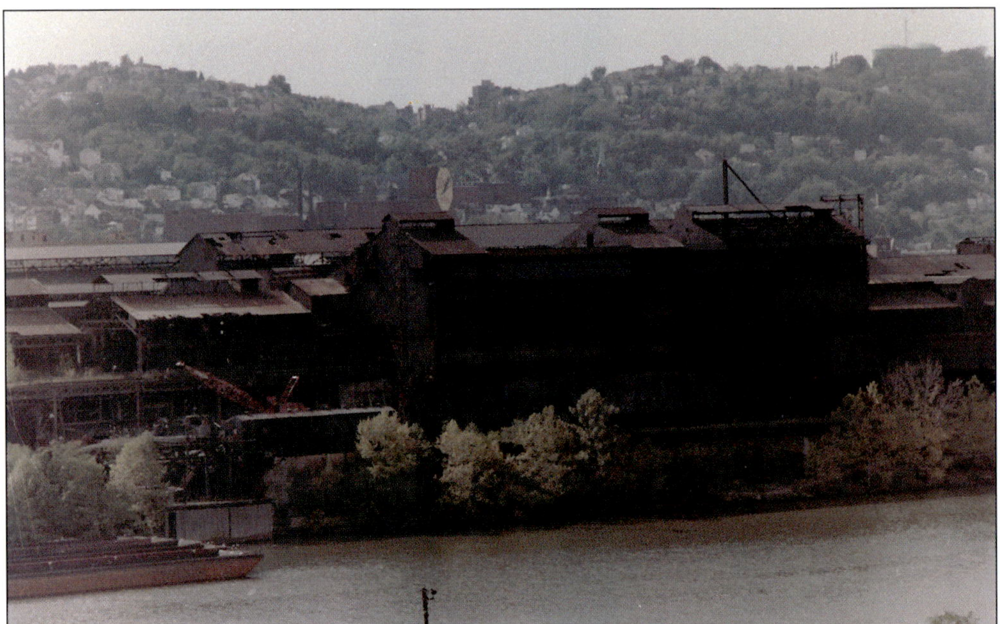

Just to the west (right) of the previous photograph, one can see the yellow-faced Duquesne brewery clock peeking over the tops of the mill buildings; it said, "Have a Duke!"

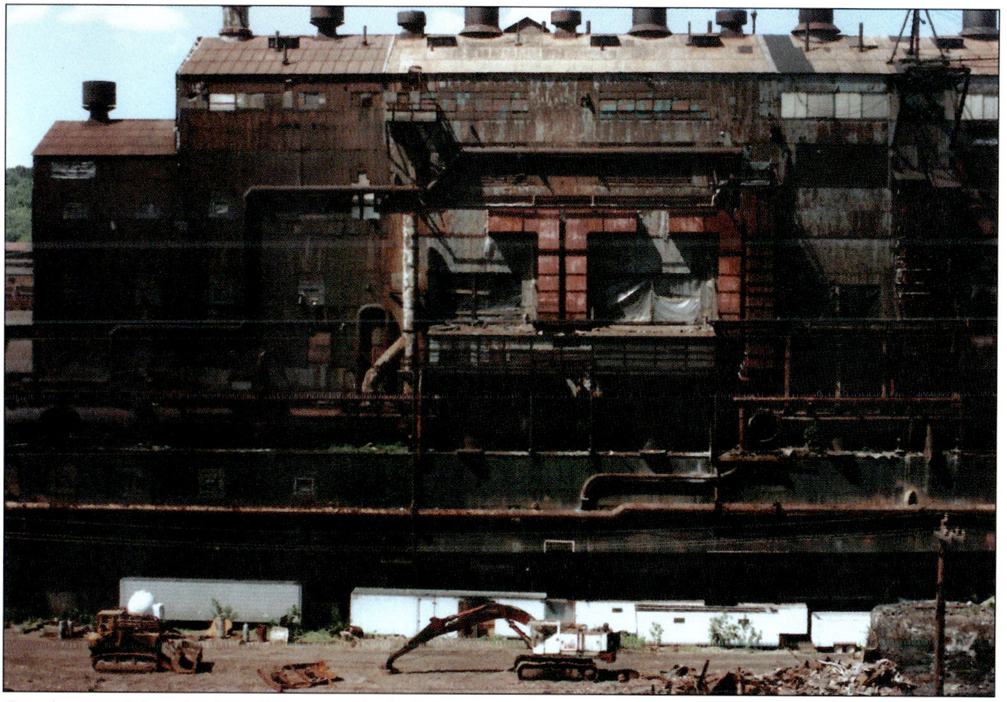

On the north bank of the Monongahela River was part of Jones and Laughlin's Eliza Blast Furnace. Seen here on a fine summer day, the machines that will demolish this building wait beneath it in the foreground.

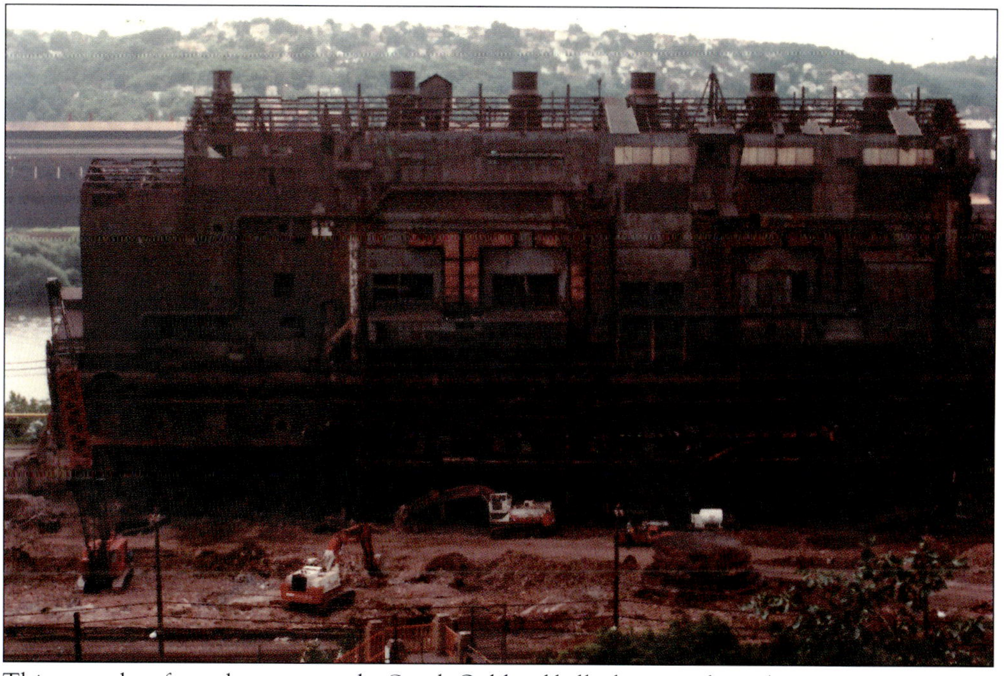

This was taken from the stairs on the South Oakland hillside across from the Jones and Laughlin Eliza Blast Furnace site along Second Avenue. The furnaces are gone, and this building was behind them. However, the building's roof has already been taken off.

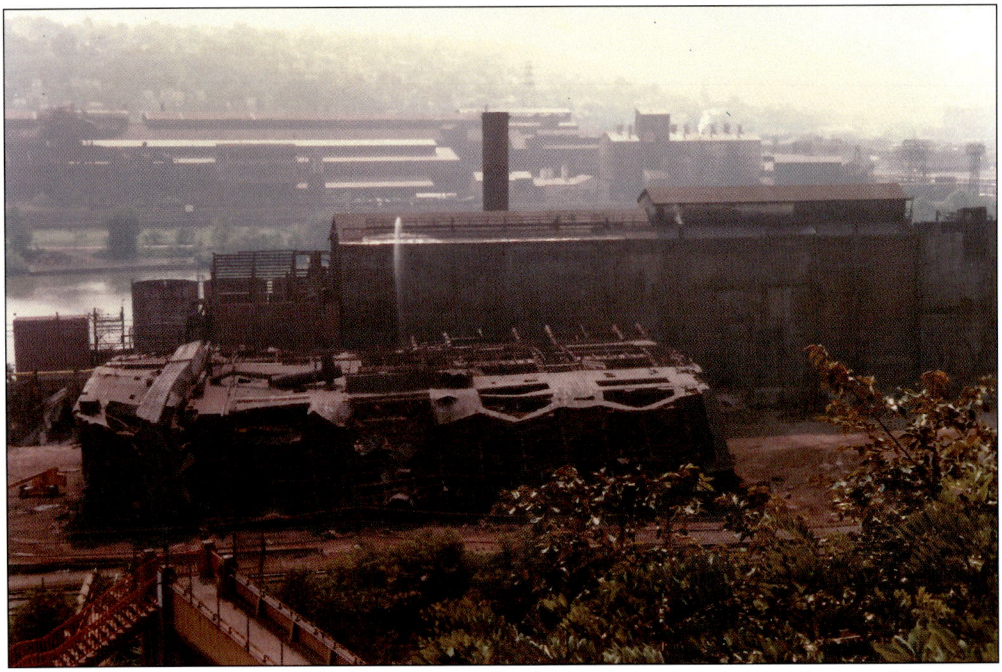

The building now lies on its side, like a great fallen beast. One can see the pedestrian foot bridge that workers used to go from their homes on the hillside to the mill and back. There is water spouting up behind the building.

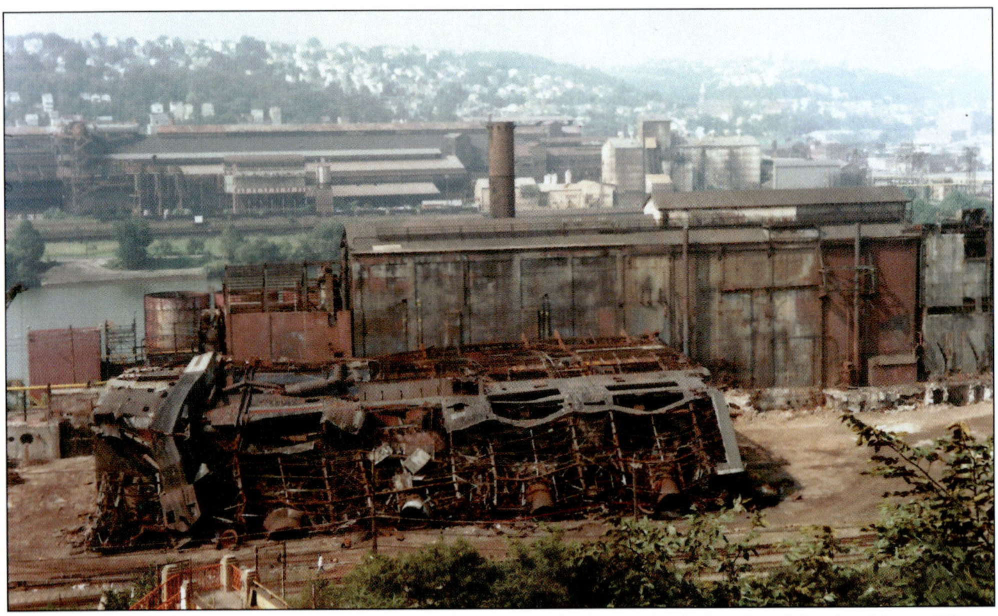

The next day was sunnier. Note the twisted elevator shaft on the left side of the building. There is also a man in a white shirt walking to the left, having just come from behind the trees.

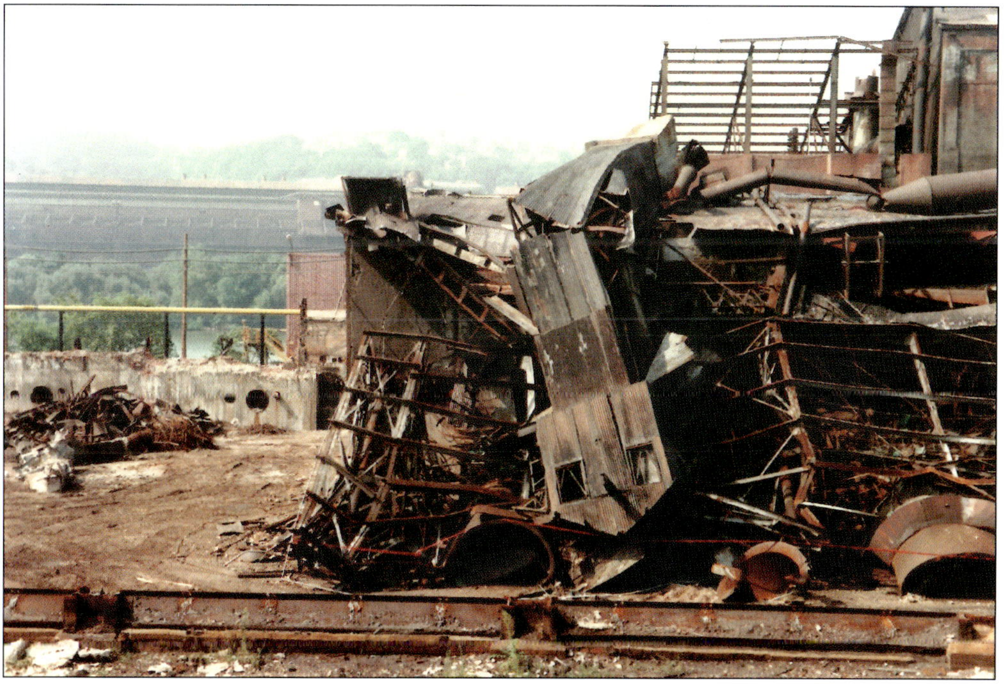

Here is a detail shot taken from the pedestrian bridge over the parkway (Interstate 376). Red caution tape has been wrapped around the building. This is the north side and the east end of the building.

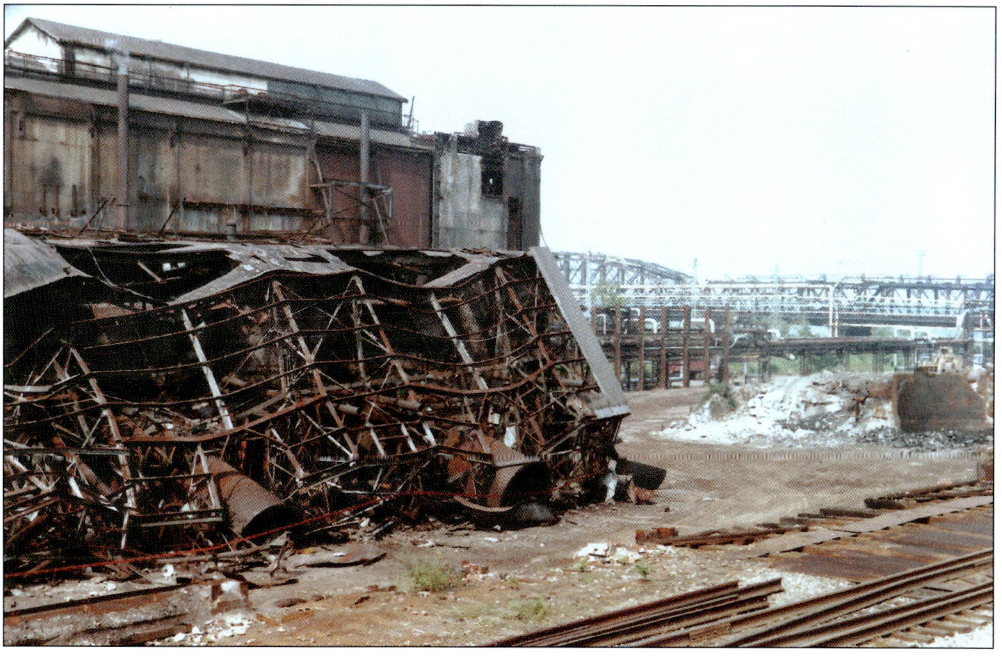

Moving to the right, one can see the west end of the building with the hot metal bridge cutting across to the right. Some old steel rails, possibly from the Monongahela Connecting Railroad, are in the foreground.

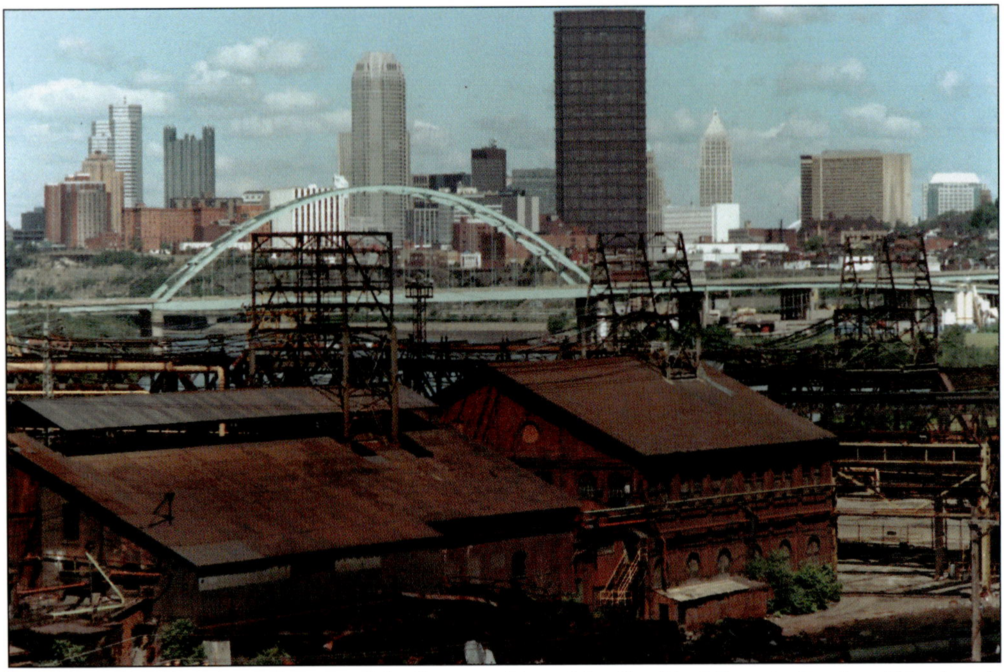

Looking northwest across the tops of the Jones and Laughlin Steel buildings is the city of Pittsburgh skyline. The tall, dark structure at center right is the US Steel Building. The Birmingham Bridge is in the center of the photograph. It connects East Cason Street on the South Side with Fifth and Forbes Avenues going uptown and to Oakland.

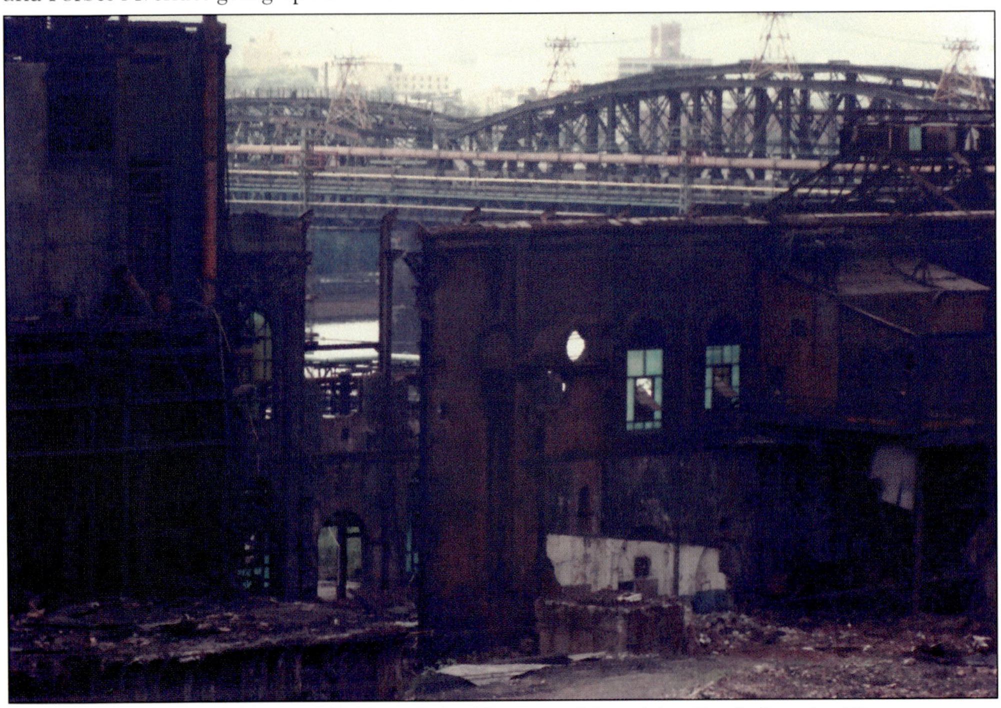

In this photograph of the partially demolished Jones and Laughlin Steel plant buildings, one can see the hot metal bridge running along the top.

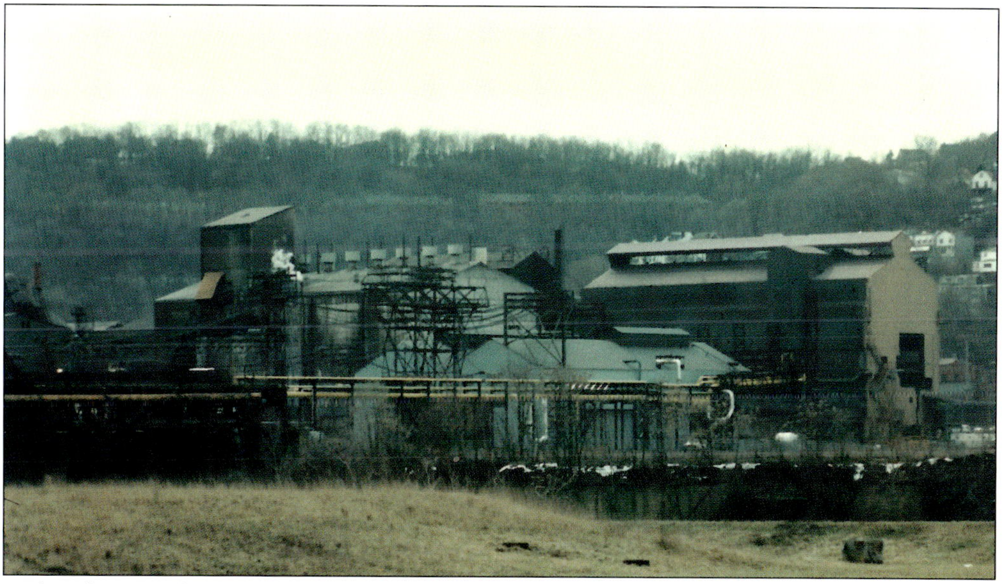

In this photograph is Jones and Laughlin's hot metal bridge. It carries a yellow natural gas pipeline over the Monongahela River. The cleared land in the foreground is now the site of Thermo Fisher Scientific, Inc., created by the merger of Thermo Electron with Fisher Scientific in 2006. Across the river, one can see the tan side of the electric arc furnace building.

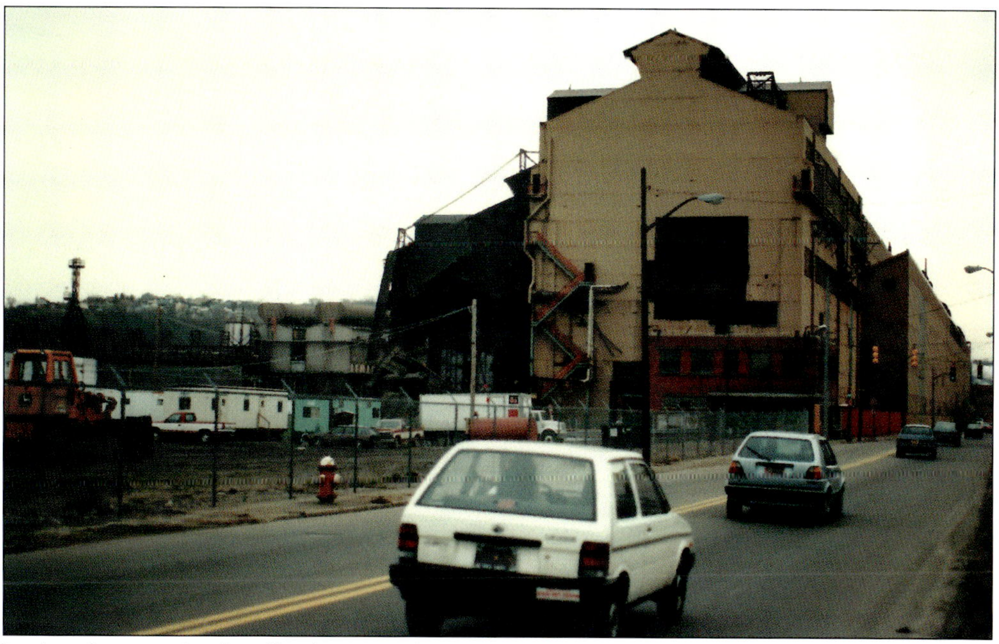

In this photograph of traffic along South Carson Street, one can see clearly the close proximity of the plant to the street. The cleared space to the left used to be filled with other Jones and Laughlin Steel buildings. The tan-sided structure was the electric arc furnace building where steel scrap was melted down.

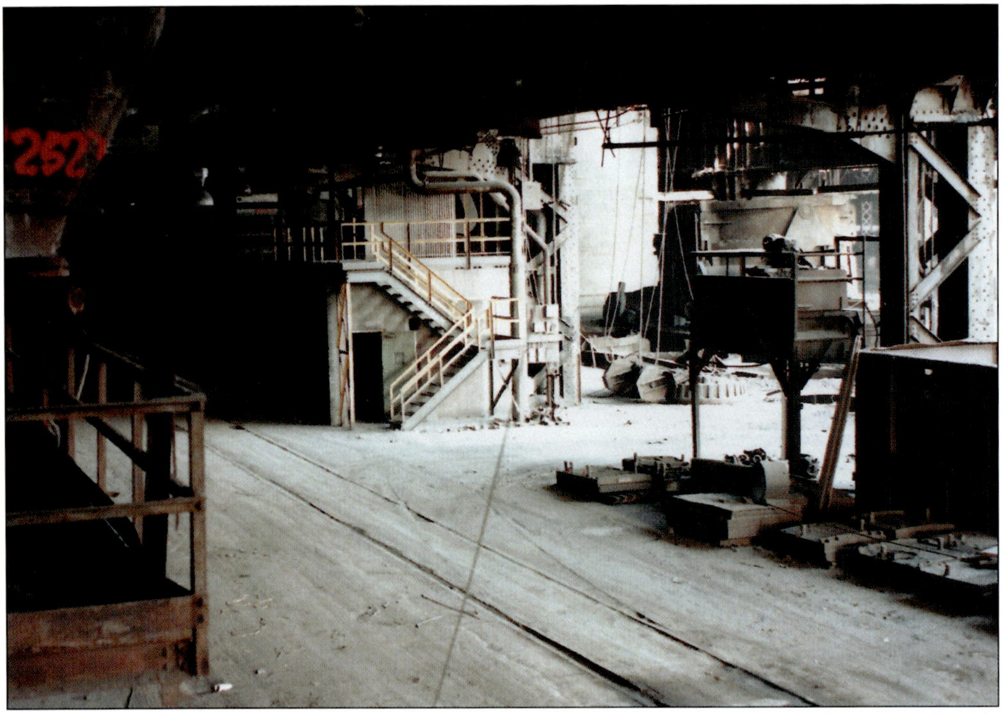

Pictured here is the inside of the electric arc furnace building of Jones and Laughlin's South Side works. Note the train tracks in the floor. This was taken in 1989.

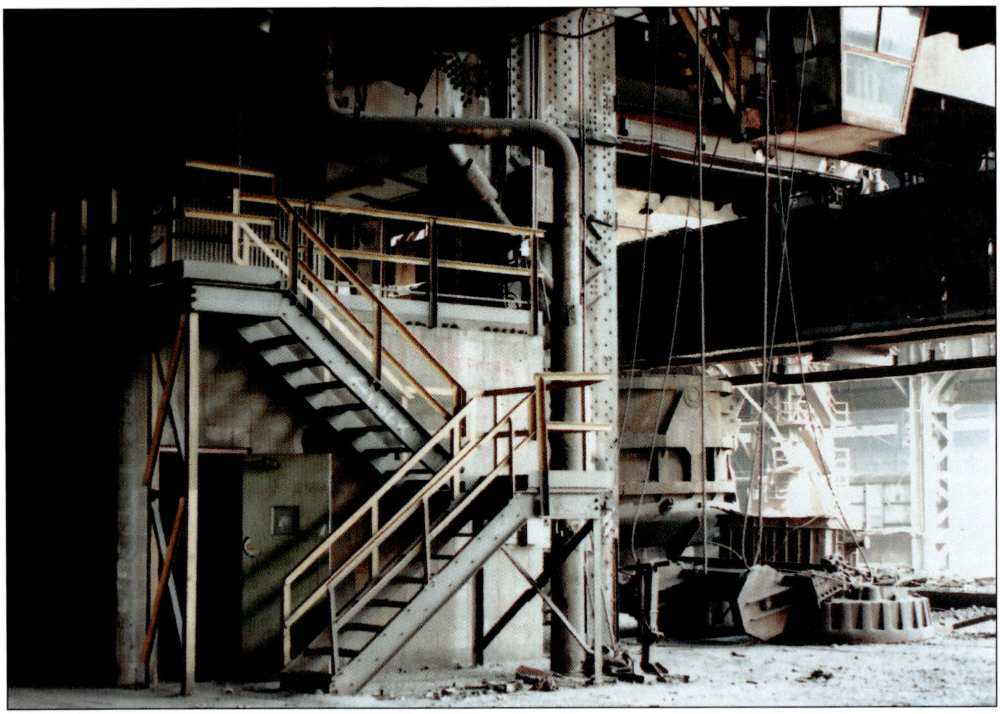

Here is a close-up of the previous image. Note the electromagnet on the shop floor to the right. Next to it on the left is a large pouring ladle.

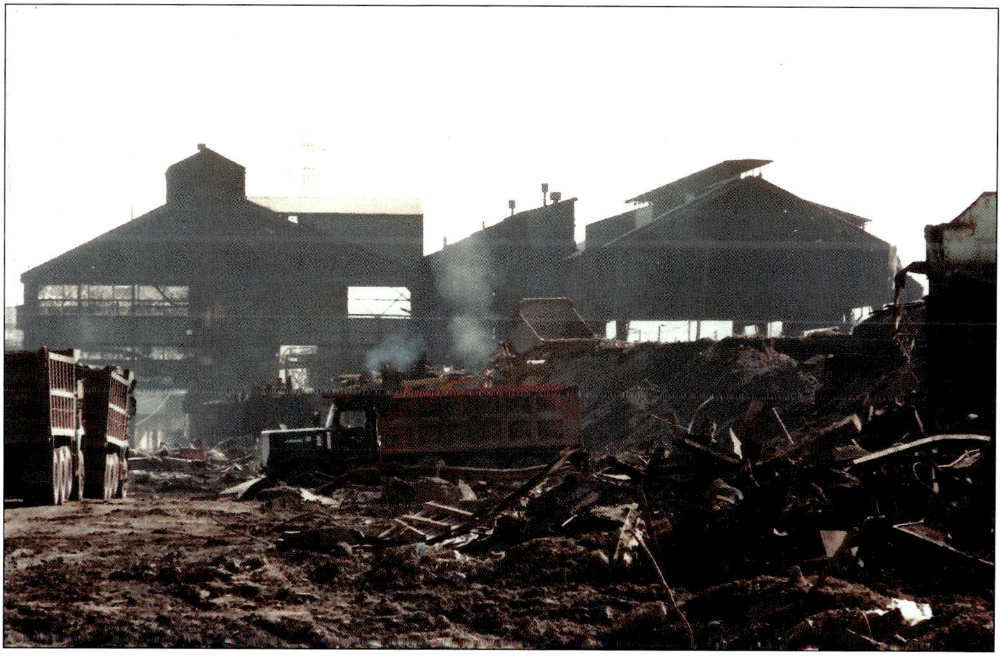

Dump trucks are being filled with debris from demolished steel plant buildings at Jones and Laughlin's South Side works. In 1989, they are clearing away the last remnants of a century of hard work, grime, and pride.

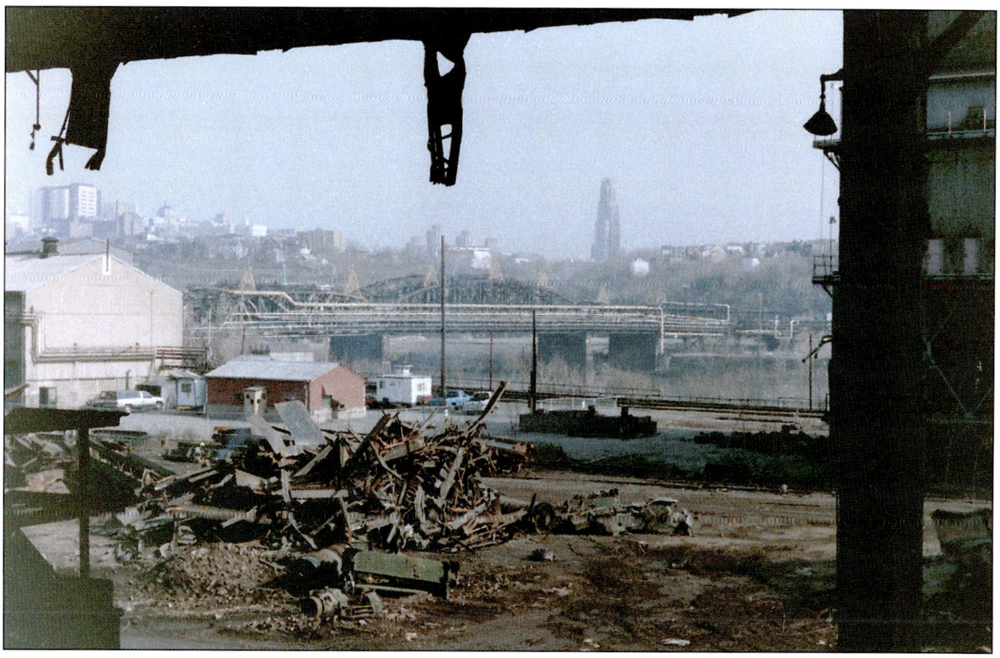

Taken from Jones and Laughlin's South Side works, this photograph shows the hot metal bridge on the left and the University of Pittsburgh's Cathedral of Learning in Oakland in the distance. Many people working in the mill wished they could afford to go to college and get out of the hot, dirty mills, and many college kids wondered if they could make it as a steel worker—such is human nature.